Andreas Prahs

A Gradient Crystal Plasticity Theory Based on an Extended Energy Balance

Schriftenreihe
Kontinuumsmechanik im Maschinenbau
Band 17

Karlsruher Institut für Technologie (KIT)
Institut für Technische Mechanik
Bereich Kontinuumsmechanik

Hrsg. Prof. Dr.-Ing. habil. Thomas Böhlke

Eine Übersicht aller bisher in dieser Schriftenreihe erschienenen Bände finden Sie am Ende des Buchs.

A Gradient Crystal Plasticity Theory Based on an Extended Energy Balance

by
Andreas Prahs

KIT Scientific Publishing

Karlsruher Institut für Technologie
Institut für Technische Mechanik
Bereich Kontinuumsmechanik

A Gradient Crystal Plasticity Theory
Based on an Extended Energy Balance

Zur Erlangung des akademischen Grades eines Doktor-Ingenieurs
von der KIT-Fakultät für Maschinenbau des
Karlsruher Instituts für Technologie (KIT) genehmigte Dissertation

von M.Sc. Andreas Prahs

Tag der mündlichen Prüfung: 05. November 2019
Hauptreferent: Prof. Dr.-Ing. Thomas Böhlke
Korreferenten: Prof. Dr. rer. nat. Bob Svendsen
Prof. Dr. Erica Lilleodden

Impressum

KIT Scientific Publishing

Karlsruher Institut für Technologie (KIT)
KIT Scientific Publishing
Straße am Forum 2
D-76131 Karlsruhe

KIT Scientific Publishing is a registered trademark
of Karlsruhe Institute of Technology.
Reprint using the book cover is not allowed.

www.ksp.kit.edu

Print on Demand 2020 – Gedruckt auf FSC-zertifiziertem Papier

ISSN 2192-693X
ISBN 978-3-7315-1025-3
DOI 10.5445/KSP/1000117916

Zusammenfassung

Plastische Deformationen induzieren in kleinskaligen metallischen Proben Größeneffekte. Um das zugehörige, nichtlokale Materialverhalten zu modellieren, werden häufig verallgemeinerte Kontinuumsmodelle verwendet. Diese berücksichtigen, im Gegensatz zum klassischen Cauchy-Boltzmann Kontinuum, zusätzliche Freiheitsgrade und führen intrinsisch eine Längenskala ein.

Die vorliegende Thesis gibt zunächst einen Überblick über verschiedene Methoden zur Herleitung erweiterter Kontinuumsmodelle. Hierbei werden die grundlegenden Annahmen und Einschränkungen der jeweiligen Methode dargestellt. Zudem werden unterschiedliche Kontinuumsmodelle sowie deren Wechselbeziehungen untereinander ausführlich diskutiert.

Basierend auf diesen Erkenntnissen wird eine selbstkonsistente Theorie hergeleitet, indem die Invarianz einer erweiterten Energiebilanz bezüglich der euklidischen Transformation betrachtet wird. Die Erweiterung der Energiebilanz basiert auf einem zusätzlichen Freiheitsgrad, für den die Anwendung des Coleman-Noll-Verfahrens eine thermodynamisch konsistente, nichtlokale Fließregel liefert. Die Äquivalenz zwischen dem betrachteten Ansatz und einem erweiterten Prinzip der virtuellen Leistung wird dargestellt. Eine Gradientenplastizitätstheorie wird im Rahmen kleiner Deformationen und Einfachgleitung aufgestellt, indem die plastische Abgleitung als zusätzlicher Freiheitsgrad betrachtet wird. Die Partitionierung der nichtlokalen Fließregel für die plastische Abgleitung liefert eine lokale Fließregel sowie eine partielle Differentialgleichung, die häufig als Mikrokraftbilanz bezeichnet wird.

Im Rahmen der erweiterten Energiebilanz wird eine materielle singuläre Fläche berücksichtigt, um das mechanische Verhalten an der Korngrenze abzubilden. Es werden thermodynamisch konsistente Fließregeln für die plastische Gleitung an der Korngrenze hergeleitet. Hierbei wird eine Klassifizierung der erhaltenen Fließregeln vorgenommen.

Abschließend wird die hergeleitete Theorie auf ein zwei- sowie ein dreiphasiges Laminat angewendet. Während das Zweiphasenlaminat aus einer elastoplastischen und einer elastischen Phase besteht, weist das Dreiphasenlaminat zwei benachbarte elastoplastische Phasen und eine elastische Phase auf. Die elastoplastische Phase repräsentiert ein Korn mit einem aktiven Gleitsystem. Die Auswirkungen der internen Längenskala und der Kornbreite auf die Verteilung der plastischen Abgleitung im Korn und auf der Korngrenze werden untersucht. Hierbei werden analytische Lösungen für die Verteilung der plastischen Abgleitung sowohl für das zwei- als auch für das dreiphasige Laminatmaterial diskutiert.

Summary

Plastic deformations induce size effects in small-scale metallic samples. Generalized continuum models are often used to model the associated nonlocal material behavior. In contrast to the classical Cauchy-Boltzmann continuum, these models consider additional degrees of freedom and intrinsically introduce a length scale.

This thesis first gives an overview of different methods for the derivation of extended continuum models. The basic assumptions and limitations of each method are presented. In addition, different continuum models and their interrelations are discussed in detail.

Based on these findings, a self-consistent theory is derived by considering the invariance of an extended energy balance with respect to the Euclidean transformation. The extension of the energy balance is based on an additional degree of freedom for which the application of the Coleman-Noll method provides a thermodynamically consistent, nonlocal flow rule. The equivalence between the approach considered and an extended principle of virtual power is presented. A gradient plasticity theory is established in the context of small deformations and single slip by considering the plastic slip as an additional degree of freedom. The partitioning of the nonlocal flow rule for the plastic slip provides a local flow rule as well as a partial differential equation, which is often referred to as the micro-force balance.

Within the framework of an extended energy balance, a material singular surface is considered in order to model the mechanical behavior at the grain boundary. Thermodynamically consistent flow rules for the plastic slip at the grain boundary are derived. Thereby, a classification of the obtained flow rules is carried out.

Finally, the derived theory is applied to a two-phase and a three-phase laminate. While the two-phase laminate consists of an elastoplastic and an elastic phase, the three-phase laminate has two adjacent elastoplastic phases and an elastic phase. The elastoplastic phase represents a grain with an active slip system. The effects of internal length scale and grain width on the distribution of the plastic slip in the grain and on the grain boundary are examined. In this context, analytical solutions for the distribution of the plastic slip for both the two-phase and three-phase laminate material are discussed.

Acknowledgments

First of all, I would like to thank Prof. Dr.-Ing. Thomas Böhlke for giving me the opportunity to work, teach, and do research at the Institute of Engineering Mechanics (ITM), as well as for the supervision of my doctoral thesis. Among others, his tremendous knowledge in mechanics along with his enthusiasm for this topic are responsible for my deep interest in this field. I appreciated our discussions, his trust in me and my work, as well as his support and guidance. Furthermore, I am also grateful for the opportunity to participate in many national and international conferences.

I would also like to express my gratitude to Prof. Dr. rer. nat. Bob Svendsen and Prof. Dr. Erica Lilleodden for co-supervising my doctoral thesis. Our discussions resulted in improvements of my thesis. Especially, Prof. Svendsen's publications (including the footnotes contained) were groundbreaking for my own research. In addition, I want to thank Prof. Dr.-Ing. Heilmaier for chairing the thesis committee.

Furthermore, I thank Jun.-Prof. Dr. rer. nat. Matti Schneider for interesting and enjoyable discussions about mechanics and its observers.

Thanks to my great colleagues, I really enjoyed my time at the ITM. Our conversations during lunch or coffee breaks were a welcome distraction. The social events, organized by Hannes and Alex, were always a lot of fun. I greatly appreciated the enjoyable discussions with my colleagues Ulrich, Tim, Olga, and Kai from the Dynamics Department at ITM as well as their motivation. Many thanks also to my former room mates Róbert, Christian, Patrick and Jannick for refreshing discussions. The preparation of the tutorials in *Technische Mechanik* would not have been the same without my discussions with Malte, Johannes R., Johannes G., and Róbert. I enjoyed the discussions on my research topic with Eric, Mauricio, Daniel, Johannes R., and Hannes. Moreover, I am grateful for all the discussions with Tom on research, teaching, organization, Linux, and other topics; I have learned a lot. Special thanks go to Ute and Helga, who always helped me in many organizational and other concerns and always had an open ear.

I am thankful for the friendship and motivation of my friends Martin, Christian, Katharina G., and Maximilian.

A big thanks also go to my parents-in-law Michael and Emilia, as well as to Michelle and Max for their great and untiring motivation during my doctoral thesis and my studies.

Special thanks go to my sister Katharina, who has always encouraged and motivated me in all matters. I would like to express my deepest gratitude to my parents Rudolf and Sigrun, who raised me with patience and love, always believed in me, and laid the foundation upon which my education is built.

Wholeheartedly, I want to thank my dear wife Loredana for her love and confidence in me and my work. She stood always and in everything by my side. I could always count on her support, not only in the field of continuum mechanics.

The support of the German Research Foundation (DFG) in the project 'Dislocation based Gradient Plasticity Theory' of the DFG Research Group 1650 'Dislocation based Plasticity' under Grant BO 1466/5 is gratefully acknowledged.

Karlsruhe, August 2020 Andreas Prahs

Contents

Chapter 1

Introduction

1.1 Motivation

The plastic deformation of metals is a crucial characteristic with respect to their processing. Metals are used for tools since thousands of years. However, it was not until the beginning of the 20th century that theories were developed describing the mechanisms of plastic deformation of metals. Amongst others, seminal works concerning this topic are by Mises (1928), Orowan (1934a;b;c; 1935a;b), Polanyi (1934) and Taylor (1934). Since metals are crystals, they consist, theoretically, of a periodic arrangement of unit cells. However, this lattice structure is not perfectly periodic as it contains defects. A line defect in this lattice is referred to as dislocation. The movement of dislocations in a metallic specimen under applied load causes its plastic deformation. In this context, the movement of the dislocations is not arbitrary. It occurs with respect to so-called slip directions and slip planes that are characteristic for a specific class of crystals.

Molecular dynamic (MD) simulations as well as discrete dislocation dynamics (DDD) simulations are used to investigate the basic mechanisms of dislocation motion such as the interaction of dislocations with each other or with obstacles. Both methods, however, require a comparably high numerical effort and are, therefore, time consuming. Consequently, they are not applied if grain aggregates are considered. If the plastic deformation of oligo- or polycrystals is of interest, continuum mechanical approaches are commonly applied. In the continuum mechanical context, the individual, discrete dislocations are not resolved separately. Instead, either a density of dislocations or the plastic slip is considered. This makes the continuous description numerically more efficient than, e.g., the MD or DDD calculations. Regarding small strain theories, it is common to presume that the plastic strain tensor is given by a linear combination of the plastic slips and the corresponding Schmid tensors that specify the slip directions and the slip planes. If a constitutive law for the evolution of the plastic slips can be established, it is possible to describe the plastic deformation of polycrystals. This approach is denoted as classical crystal plasticity theory. It yields reasonable results, if the considered specimen exhibits a sufficient size, e.g., in the context of deep drawing of sheet metals. However, if the specimen size or the grain size under consideration decreases to the scale of a few micrometers, the results obtained are not in line with corresponding experiments. Size effects such as an increase in the total yield strength are observed but are not reproducible within the framework of classical crystal plasticity. They are related to the pile-up of geometrically necessary dislocations (GNDs) at grain boundaries (GBs) or other obstacles.

The crystal lattice is locally compressed or stretched due to the presence of dislocations. Therefore, each dislocation carries a stress field with it. Consequently, pile-ups of GNDs are associated with inhomogeneities of stress and therefore with strain gradients. To take this into account, gradient plasticity theories were developed. Some of these theories account for the plastic slips at the individual slip systems as additional degrees of freedom (DOFs). Thus, the stored free energy depends on the gradients of the plastic slips. If additional DOFs are accounted for each material point, the continuum under consideration is not referred to as classical but as generalized or extended continuum. A prominent example for an extended continuum is the Cosserat continuum. It considers three additional rotational DOFs for each material point. In general, the influence of the supplementary DOFs on the common balance equations has to be discussed. In this context, mainly two questions arise.

- Which balance equations of the classical continuum will be preserved in their standard form?
- Are there additional balance equations associated with the additional DOF?

Both questions can be answered by deriving the balance equations for an extended continuum. In order to obtain a closed theoretical framework, constitutive equations as well as flow rules for additional DOFs are needed. For the derivation of balance and constitutive equations, several methods are applicable and known from literature. However, the interrelation between these methods as well as their fundamental assumptions and their limitations are commonly not discussed. In fact, however, the restrictions associated with each method are of particular interest for the development of a thermodynamically consistent, extended plasticity theory. After these questions have been clarified, the influence of geometrical properties of the GB, e.g., its mean curvature, on the balance equations can be discussed.

1.2 Research Objectives and Originality

Research objectives The *main objective* of this thesis is the thermodynamically consistent development of extended plasticity models describing scale effects. Thereby, materials with a crystalline microstructure, such as metals, are considered. To this end, a generalization of the classical Cauchy continuum is used that takes into account additional degrees of freedom (DOFs). The treatment of this topic by means of either variational principles or an extended principle of virtual power (PoVP) is widely spread. A third approach, the consideration of invariance properties of an extended energy balance, is underrepresented in modern literature. While the exploitation of a variational principle is straightforward, it does not take into account the second law of thermodynamics. Instead, a dissipation potential is considered. The description of the problem by means of an extended PoVP is both elegant and effective to apply. However, the balance of internal energy is commonly introduced in addition, to account for supplementary contributions in the dissipation inequality by means of the Legendre transformation. For this purpose, in addition to the virtual rates used by the PoVP, actual rates are introduced. This gives rise to some redundancy. In contrast to this, the invariance considerations of an extended energy balance naturally yield a balance of internal energy which accounts for additional contributions. Consequently, the last approach is self-consistent. However, it is not possible to obtain additional field equations by means of invariance considerations of an extended energy balance. This raises the question of the relationship between the results of an extended PoVP and the invariance of an extended energy balance.

The *second objective* of the work at hand is the development of an extended plasticity model that takes into account the mechanical behavior at the GB. As known from experiments on oligo- or polycrystals, the overall mechanical behavior is significantly influenced by GBs. They constitute a resistance against the movement of dislocations and result in a pile-up at the GB. These pile-ups are associated with strain gradients.

Finally, the *third objective* is to apply the derived framework to a slip gradient crystal plasticity theory in the context of small deformations. The qualitative characteristics of the distribution of the plastic slip are discussed by means of analytic solutions. In this context, single slip within a two- and a three-phase laminate material is considered. The two-phase laminate material is used to discuss the mechanical behavior within the bulk material. In addition, the three-phase laminate yields results with respect to the mechanical behavior at the GB.

Originality of this thesis The following novelties are obtained.

- The invariance considerations of an extended energy balance allow for a self-consistent development of extended plasticity models. In this context, an extended energy balance and an extended PoVP are compared with respect to regular points. Thereby, the notion of the micro-force balance is discussed. In contrast to common approaches using an extended PoVP, associated micro-inertia terms are taken into account, here.

- An extended energy balance that accounts for a material singular surface as a model for a GB is considered. The mean curvature of the GB is explicitly accounted for by its corresponding balance of mass. Moreover, flow rules for a rate-independent, nonlinear behavior of the plastic slip at the GB are obtained. Boundary conditions obtained by an extended PoVP constitute a special case of the rate-independent behavior.
- Considering a small strain slip gradient crystal plasticity, an analytical solution for the plastic slip is provided for a three-phase laminate material with respect to single slip. This enables to discuss the sensitivity of the plastic slip at the GB by analytical means with respect to a change of the material parameters. In this context, the influence of the defect parameter, which is associated with an initial dislocation density, is considered. Regarding adjacent grains with geometrically coherent slip systems, slip gradients are obtained at the GB based solely on a jump of the defect parameter and, thus, the initial dislocation density.

1.3 State of the Art[1]

Basic ideas of extended continua An extensive review on extended continua is given in the subsequent chapter. Here, the development of extended continua throughout the 20th and the early 21st century is briefly outlined and the corresponding seminal publications are given. Classical continuum mechanics considers a body as a set of undeformable material points. Each material point of such a Boltzmann continuum

[1] Most of the content of this section is taken directly from the articles Prahs and Böhlke (2019b;a). Minor linguistic changes and abbreviations have been made.

exhibits three degrees of freedom (DOFs) describing its displacement, cf. Hellinger (1913, p. 606) and Eugster and dell'Isola (2017; 2018a;b). The Boltzmann continuum, cf. Vardoulakis (2019, p. 1), is also referred to as Cauchy continuum as described in Maugin (2017, p. 3), cf. also dell'Isola et al. (2015); Giorgio (2016); Rahali et al. (2015). Extended continuum models account for the underlying microstructure of the material by introducing additional, internal DOFs. One of the first suggested extended continuum models is the so-called Cosserat continuum, cf. Cosserat and Cosserat (1909). It allows for the orientation of a material point. Thus, each material point is supplemented by three rotational DOFs in addition. Many authors addressed this topic in the mid of the 20th century focusing on generalizations or extensions to the Cosserat continuum, cf. Ericksen (1961); Eringen (1964; 1968); Germain (1973); Green and Rivlin (1964a); Mindlin (1964). Especially the micromorphic continuum according to Eringen and Suhubi (1964) can be considered as a direct generalization of the Cosserat continuum. It treats each material point as a micro-continuum. Consequently, a micro-deformation tensor associated with each material point is introduced in Eringen and Suhubi (1964). Conceptually similar is the consideration of a micro-medium as discussed by Mindlin (1964). Continua that account for couple stresses are discussed by Toupin (1962; 1964). Velocity gradients of higher order or multipolar displacements are introduced in the context of extended continua by Green and Rivlin (1964c;a). An extensive overview of generalized continua is given in Capriz (1989) and Neff et al. (2014). Further applications of extended continua are to be found in the context of liquid crystals, cf. Ericksen (1961); Leslie (1968), continuum theory of dislocations, cf. Fox (1966; 1968), nonlocal plasticity, cf. Peerlings et al. (2004); Placidi (2016), nonlocal damage, cf. Germain et al. (2007); Placidi and Barchiesi (2018); Placidi et al. (2018a;b) and nonlocal diffusion, cf. Ubachs et al. (2004).

Methods to obtain field equations of an extended continuum Additional DOFs are associated with the corresponding equations of motion relating the kinematic of the DOFs to the underlying forces, cf. remark in Maugin (2015). Consequently, the total energy describing the system is supplemented by contributions related to the additionally introduced DOFs. The comparison of a mathematical pendulum with a double pendulum serves as an illustrative example with respect to discrete systems, cf. Landau and Lifshitz (1969). In the continuum mechanical context, several approaches exist to derive or motivate additional balance equations associated with additional DOFs. An overview is given in the review paper of Mariano (2016) or others, e.g., Misra et al. (2017); Placidi et al. (2017). In a nondissipative context, Hamilton's principle of least action is a suitable method for the derivation of associated field equations, cf. Hellinger (1913). It can be considered as the predecessor of many other variational principles. An application to continua with a microstructure based on elastic micro-trusses is given by Seppecher et al. (2011). However, this is getting more involved for dissipative systems, cf. Planck (1960, p. 81). Closely related to variational principles is the principle of virtual power. Its classical formulation can be supplemented by additional work terms accounting for the virtual power of additional DOFs, cf. Forest (2009). According to Mariano (2016), a drawback of an extended principle of virtual power is that quantities, such as the stress and micro-stress tensor, are presumed. Another approach is to consider the invariance properties of an extended energy balance with respect to a superimposed rigid-body motion. Additionally, the 'tetrahedron' argument, cf., e.g., dell'Isola et al. (2016), is applied to prove the existence of, e.g., the stress and micro-stress tensor. This approach is often referred to as Green-Naghdi-Rivlin (GNR) theorem, cf. Marsden and Hughes (1994); Maugin (1980). Its first application can be found in Green and Rivlin (1964b). With this method, however, it is difficult to obtain additional field equations, as already noted by Planck (1960). In Germain (1973, p. 574) it is stated that invariance considerations of an extended energy balance do not lead to the same field equations as obtained by an extended principle of virtual power. Maugin (1980) confirms this issue referring to the seven parameter invariance that is commonly applied to the energy balance in this context. The number of field equations obtained by the energy balance is less compared to the number obtained by Hamilton's principle, cf. Planck (1960), or the principle of virtual power, cf. Maugin (1980, p. 63). This topic is recaptured by Yavari and Marsden (2009, p. 10). They show that an extended energy balance only leads to modifications of the common balance equations if the ambient space is chosen Euclidean rather than a Riemannian manifold. However, it is not possible to obtain additional field equations. In Yavari and Marsden (2009), the extensions to the energy balance, which is discussed in the context of the GNR-theorem, are due to additional vectorial DOFs. As discussed in Svendsen (2001a, footnote 2), it is regardless whether the invariance of the energy balance is considered with respect to a superimposed rigid-body motion or a change of observer. Both approaches yield the same balance equations.[2] In fact, some authors state the invariance of an extended energy balance with respect to a change of observer, e.g., Capriz et al. (1982). Invariance considerations of an extended energy balance are discussed by Svendsen (2001b) in the context of a fibre bundle framework, thereby extending the framework of Capriz (1989). Svendsen (2011) generalized the treatment of Capriz (1989) from simple materials to first-order gradient continua. Moreover, the covariance of an extended energy balance with respect to a spatial and a microstructural diffeomorphism is considered in Yavari and Marsden (2009). This procedure is applied to a classical Boltzmann continuum in Marsden and Hughes (1994, pp. 165-167). However, they explicitly emphasize that the stress vector transforms objectively, irrespective of the underlying material behavior, if the considered spatial diffeomorphism describes a rigid deformation. Regarding a generic spatial diffeomorphism, an objective transformation of the stress vector is postulated only in the

[2] In the context of material theory, invariance considerations with respect to a change of observer or a superimposed rigid-body motion are denoted as PMO or PISM, respectively. In contrast to the PMO, the PISM is not always valid, cf. Krawietz (1986, p. 161), Svendsen and Bertram (1999) and Svendsen (2001a).

purely elastic case, cf. Marsden and Hughes (1994, p. 163). Consequently, the consideration of dissipative processes by means of this framework is quite involved. This limitation to purely elastic material behavior can be seen as the most critical point of this approach. Following Truesdell and Toupin (1960, p. 529), a clear separation between balance equations and constitutive equations has to be drawn. This arises from the demand that balance equations should be of generic nature, valid for all materials. Hence, as stated in Truesdell and Toupin (1960), constitutive laws cannot be obtained from balance equations. Contrarily, the approach of Marsden and Hughes (1994) and Yavari and Marsden (2009) yields a constitutive equation for the stress tensor, denoted as Doyle-Ericksen formula.

Extended continua in the context of crystalline materials Regarding crystalline materials, the pile-up of dislocations at obstacles, such as GBs, is associated with inhomogeneous plastic slips, cf. Aifantis et al. (2006); Bayerschen et al. (2015). Thus, interactions between dislocations and obstacles play an important role with respect to nonlocal mechanical behavior. Such interactions always involve an internal length scale, e.g., the intermediate distance between two dislocation lines. A coarse grained measure for such an internal length scale is given, for instance, by the dislocation density stored in the material. The smaller the size of a considered specimen, the more pronounced are so-called size effects due to interactions at the internal length scale. A conspicuous example is given by the Hall-Petch effect, cf. Hall (1951) and Petch (1953), which has been experimentally investigated for several metals, cf. Armstrong et al. (1962). To account for such phenomena, the plastic slip is introduced as additional degree of freedom (DOF) at each material point. The internal length scale is commonly associated with gradient terms of the plastic slip. Continua that consider additional DOFs are commonly denoted as extended, or generalized continua, cf. Forest (2009).

Grain boundaries as obstacles against dislocation movements Considering oligo- or polycrystals, the nonlocal mechanical behavior can be traced back to the presence of GBs and their resistance against dislocation movement, cf. Aifantis and Willis (2005). A continuum mechanical treatment of GBs is given by Gurtin and Needleman (2005) regarding small deformations. The influence of the misorientation of adjacent grains and of the GB is discussed in detail by Gurtin (2008) and Gottschalk et al. (2016). Moreover, an extensive overview of slip transmission criteria at the GB is presented by Bayerschen et al. (2016a). GBs can be modeled as material singular surfaces in a thermodynamical context. Regarding gradient plasticity theories based on an extended principle of virtual power or on a variational principle, additional terms are introduced on the singular surface, cf. Gurtin (2008); Aifantis et al. (2006). Consequently, a specific free energy accounts for mechanisms at the GB such as a constant slip resistance or a slip transmission-like characteristic, cf. Wulfinghoff et al. (2013); Bayerschen et al. (2016a). However, the principle of virtual power exhibits some redundancy regarding extended continua, cf. Hütter (2016). In this context, the dissipation inequality is exploited to obtain thermodynamically consistent interface conditions. By contrast, the relation between the virtual, supplementary contributions and both the energy and entropy balance is commonly not discussed. In addition, aspects that are intrinsic to the surface are often not taken into account. Particularly, the influence of the curvature of the considered material surfaces on the overall mechanical behavior is neglected from the outset. Regarding an extended energy balance with contributions on the singular surface, the transport theorem for surfaces naturally has to be applied, cf. Müller (1985). This transport theorem automatically accounts for the curvature of the singular surface, cf. Cermelli et al. (2005); Moeckel (1975); Müller (1985). The invariance of the energy balance with respect to a change of observer, cf. Marsden and Hughes (1994), yields the balance equations in regular and singular points. An entropy balance supplemented by additional terms on the singular surface provides the dissipation inequality in both regular and singular points by means of the second law of thermodynamics. Finally, the evaluation of the dissipation inequality for singular points can be used to derive constitutive equations at the GB. In this context, Triani and Cimmelli (2012) extend the interpretation of Muschik (1996) concerning the second law of thermodynamics to media involving material singular surfaces.

1.4 Outline of the Thesis

A review on extended continua is given in Chapter 2. To this end, the notion of a classical continuum is introduced first. Thereby, continuum mechanical fundamentals such as kinematics and deformation measures are briefly presented. Invariance of the energy balance with respect to a change of observer yields the balance of mass, linear and angular momentum. In addition, the entropy balance and the Clausius-Duhem inequality are given. Moreover, the principle of virtual power is formulated, and a variational principle for conservative forces and a hyperelastic material behavior is provided. After this brief review of classical continuum mechanics, the notion of an extended continuum is introduced. Different extended continuum models from literature are briefly presented. Thereby, the relationship between the different models is outlined. In this context, the invariance as well as the covariance of an extended energy balance, an extended principle of virtual power and a variational principle based on an extended Lagrangian density are considered. Finally, theories that postulate an additional balance equation associated to additional DOFs are listed.

In Chapter 3, the invariance properties of an extended energy balance are presented regarding a material volume. At first, an extension by an additional scalar-valued DOF is considered. The balance equations concerning mass, linear and angular momentum as well as the internal energy are derived extensively. A nonlocal evolution equation for an additional DOF is obtained by the exploitation of the Clausius-Duhem inequality. Finally, the set of resulting equations is compared to the equations obtained from an extended principle of virtual power. Moreover, an extension by a vectorial DOF is considered.

An extended energy balance for a material volume divided by a material singular surface is given in Chapter 4. The supplementary contributions to the energy balance are based on the plastic slip as additional DOF. For brevity, single slip is considered. The material singular surface serves as model for a GB. Mathematical preliminaries such as the divergence theorem in the presence of a material singular surface, and the transport theorem for a material surface are provided. Invariance considerations of the extended energy balance as well as the implications on the dissipation inequality are discussed. Potential relations and boundary conditions at the GB are obtained. Thereby, different possible flow rules for the plastic slip at the GB are discussed. Finally, the connection to an extended principle of virtual power is outlined.

An application of the presented theory to a slip gradient crystal plasticity with respect to small deformations and single slip is given in Chapter 5. A two-phase as well as a three-phase laminate material are considered. The analytical solutions for the distribution of the plastic slip are derived for both laminate materials. The two-phase laminate material consists of an elastoplastic and an elastic phase. Therefore it is used to consider the distribution of the plastic slip within a single crystal. The sensitivity of the solution with respect to a change of the material parameters is discussed and an upper bound for the plastic slip is given. The three-phase laminate material consists of two elastoplastic and one elastic phase. The transition between both elastoplastic phases is considered as GB. For geometrically coherent slip systems, the sensitivity of the plastic slip with respect to the defect parameter is discussed.

Finally, the main results and concluding remarks of this thesis are provided in Chapter 6.

1.5 Notation, Frequently Used Acronyms, Symbols, and Operators

Acronyms

DDD	Discrete dislocation dynamics
DOF	Degree of freedom
GB	Grain boundary
GND	Geometrically necessary dislocation
MD	Molecular dynamics
PDE	Partial differential equation
PoVP	Principle of virtual power
PoVW	Principle of virtual work

Latin letters

$a, b, A, B, \mathcal{D}, \dots$	Scalar quantities
$\boldsymbol{u}, \boldsymbol{v}, \boldsymbol{w}, \dots$	First-order tensors
$\boldsymbol{A}, \boldsymbol{B}, \boldsymbol{C}, \dots$	Second-order tensors
$\mathbb{A}, \mathbb{B}, \mathbb{C}, \dots$	Fourth-order tensors
$\tilde{A}$	Micro-inertia
$\tilde{b}$	Micro-body force
C^{++}, C^{--}	Intra-grain interaction moduli
C^{-+}	Inter-grain interaction moduli
e	Specific internal energy
E	Young's modulus
G	Shear modulus
K_g	Defect parameter
K_m	Mean curvature
$\tilde{t}$	Micro-traction force

$\tilde{v}$	Spatial velocity of additional DOF
$\mathcal{W}$	Euclidean vector space
$\boldsymbol{b}$	Body force
$\boldsymbol{d}$	Slip direction
$\boldsymbol{g}$	Temperature gradient
$\boldsymbol{n}$	Normal vector
$\boldsymbol{q}$	Heat flux vector
$\boldsymbol{t}$	Stress vector
$\boldsymbol{u}$	Displacement field
$\boldsymbol{x}$	Position vector of a material point in the current configuration
$\boldsymbol{X}$	Position vector of a material point in the reference configuration
$\boldsymbol{B}$	Left Cauchy-Green tensor
$\boldsymbol{C}$	Right Cauchy-Green tensor
$\boldsymbol{F}$	Deformation gradient
$\boldsymbol{G}$	Grain boundary Burgers tensor
$\boldsymbol{I}$	Identity tensor of second order
$\boldsymbol{K}$	Curvature tensor
$\boldsymbol{M}$	Schmid tensor
$\boldsymbol{Q}(t)$	Time-dependent isometry between two vector spaces
$\boldsymbol{R}$	Rigid body rotation tensor
$\boldsymbol{U}$	Right stretch tensor
$\boldsymbol{V}$	Left stretch tensor
$\mathbb{C}$	Stiffness tensor

Greek letters

δ	Specific dissipation
η	Specific entropy

$\tilde{\varphi}_t$	Microstructure function
Ψ	Specific free energy
Ψ_e	Specific elastic energy
Ψ_g	Specific defect energy
Ψ_h	Specific hardening energy
ρ	Mass density
ρ_0	Initial dislocation density
Θ_0	Initial hardening modulus
θ	Temperature
γ	Additional degree of freedom (e.g., plastic slip)
$\varphi_t(\boldsymbol{X})$	Deformation function
$\boldsymbol{\xi}$	Gradient stress vector
$\boldsymbol{\varepsilon}$	Infinitesimal strain tensor
$\boldsymbol{\varepsilon}^{\mathrm{e}}$	Elastic strain tensor
$\boldsymbol{\varepsilon}^{\mathrm{p}}$	Plastic strain tensor
$\boldsymbol{\sigma}$	Cauchy stress tensor
$\boldsymbol{\epsilon}$	Permutation tensor

Operators

$\boldsymbol{AB}$	Linear mapping of a second-order tensor
$\boldsymbol{A} = \mathbb{C}[\boldsymbol{B}]$	Linear mapping of a second-order tensor by a fourth-order tensor
$\boldsymbol{A} \cdot \boldsymbol{B}$	Dot product of two tensors $\boldsymbol{A}, \boldsymbol{B}$
$\boldsymbol{A} \otimes \boldsymbol{B}$	Dyadic product of two tensors $\boldsymbol{A}, \boldsymbol{B}$
$\det(\cdot)$	Determinant
$\mathrm{div}(\cdot)$	Eulerean divergence of a vector or tensor
$\mathrm{div}_{\mathcal{S}}(\cdot)$	Surface divergence of a vector or tensor
$\mathrm{Grad}(\cdot)$	Lagrangian gradient of a quantity
$\mathrm{grad}(\cdot)$	Eulerean gradient of a quantity
$\mathrm{grad}_{\mathcal{S}}(\cdot)$	Surface gradient of a quantity

∇_{n}	Planar gradient
Δ_{n}	Laplacian with respect to a planar gradient
$\mathrm{sym}(\cdot)$	Symmetric part of a quantity
$\mathrm{tr}(\cdot)$	Trace of a tensor
$(\cdot)^{\mathcal{S}}$, $(\cdot)_{\mathcal{S}}$	Quantity with respect to the surface $\mathcal{S}$
$(\cdot)^{\mathcal{V}}$, $(\cdot)_{\mathcal{V}}$	Quantity with respect to the volume $\mathcal{V}$
$(\cdot)^{+}$	Right-hand limit of a quantity
$(\cdot)^{-}$	Left-hand limit of a quantity
$(\cdot)'$	Derivative of a quantity with respect to x_1
$(\cdot)^{\mathsf{T}_\mathsf{H}}$	Major transposition, $C^{\mathsf{T}_\mathsf{H}}_{ijkl} = C_{klij}$, $\epsilon^{\mathsf{T}_\mathsf{H}}_{ijk} = \epsilon_{kji}$
$(\cdot)^{\mathsf{T}_\mathsf{L}}$	Left transposition, $C^{\mathsf{T}_\mathsf{L}}_{ijkl} = C_{jikl}$, $\epsilon^{\mathsf{T}_\mathsf{L}}_{ijk} = \epsilon_{jik}$
$(\cdot)^{\mathsf{T}_\mathsf{R}}$	Right transposition, $C^{\mathsf{T}_\mathsf{R}}_{ijkl} = C_{ijlk}$, $\epsilon^{\mathsf{T}_\mathsf{R}}_{ijk} = \epsilon_{ikj}$
$f \circ g$	Composition between two maps f and g

Chapter 2

Review on Extended Continua

Motivation Microstructured materials exhibit a nonlocal mechanical behavior, e.g., the Hall-Petch effect, cf. Hall (1951). Classical continuum models are not capable of reproducing such size effects, since they do not possess an internal length scale. Extended continuum models, such as gradient plasticity theories, overcome this drawback by introducing an internal length scale, e.g., by means of a defect energy, e.g. Bayerschen et al. (2016b). Various methods are known to derive the corresponding field equations associated with the additional DOFs, cf. the review paper of Mariano (2016). Here, the notion of a generalized continuum is introduced. To this end, the basic equations of a classical continuum are revisited first. Subsequently, extended continua from literature are presented that are based on invariance considerations of an extended energy balance, an extended principle of virtual power or work, and on an extended variational principle. Finally, some theories are listed that propose balance laws without further derivation.

2.1 The Notion of Classical and Generalized Continua

2.1.1 Classical continua

The concept of a classical continuum Classical continuum mechanics considers a body as a set of undeformable material points. Each material point exhibits three degrees of freedom (DOFs), describing its displacement, cf., e.g., Hellinger (1913, p. 606)[1]. Such a continuum is often referred to as Cauchy or Boltzmann continuum, cf. Vardoulakis (2019, p.1) or Maugin (2017, p. 3).

Kinematics and deformation The current configuration, referred to as $\mathcal{C}$, is occupied by a body that is exposed to arbitrary, external loads. An arbitrary reference configuration is denoted as $\mathcal{B}$. Within the scope of this thesis, $\mathcal{C}$ and $\mathcal{B}$ are considered to be embedded in the Euclidean space, i.e., an Euclidean ambient space is considered, cf. Yavari and Marsden (2009). A material point, identified by its position vector $\boldsymbol{X}$, is mapped from $\mathcal{B}$ to $\mathcal{C}$ by the deformation function φ_t. Regarding the current configuration, a material point is identified by its position vector $\boldsymbol{x}$. The spatial velocity field $\boldsymbol{v}$ is obtained by means of the time derivative of the deformation mapping φ_t and is calculated by

$$\boldsymbol{v}(\boldsymbol{x},t) = \left(\left. \frac{\partial \varphi_t(\boldsymbol{X})}{\partial t} \right|_{\boldsymbol{X}=\text{const.}} \right) \circ {\varphi_t}^{-1}, \quad \boldsymbol{x} = \varphi_t(\boldsymbol{X}). \tag{2.1}$$

Here, the composition between two maps, f and g, is denoted as $f \circ g$. The displacement field $\boldsymbol{u}$ is defined by

$$\boldsymbol{u}(\boldsymbol{X},t) = \varphi_t(\boldsymbol{X}) - \boldsymbol{X}. \tag{2.2}$$

[1] cf. also the note in Toupin (1964, p. 86)

The first spatial derivative of the deformation mapping is referred to as deformation gradient $\boldsymbol{F} \in \mathit{Inv}$. While $\boldsymbol{F}$ is not a gradient in the mathematical sense, cf. Marsden and Hughes (1994, p. 3), the following definition is frequently given in the continuum mechanical literature

$$\boldsymbol{F} = \mathrm{Grad}\left(\varphi_t(\boldsymbol{X})\right), \tag{2.3}$$

cf., e.g., Holzapfel (2000, p. 71). Here, $\mathrm{Grad}\left(\cdot\right)$ denotes the derivation with respect to $\boldsymbol{X}$. Since $\boldsymbol{F}$ is a so-called two-point tensor, it maps line segments from $\mathcal{B}$ to $\mathcal{C}$, cf. Bertram (2005). In this context, a line segment of the reference configuration is referred to as $\mathrm{d}\boldsymbol{X}$, and a line element of the current configuration as $\mathrm{d}\boldsymbol{x}$. Thus, the following relation holds true

$$\mathrm{d}\boldsymbol{x} = \boldsymbol{F}\,\mathrm{d}\boldsymbol{X}, \tag{2.4}$$

cf. Bertram (2005, p. 96).

Polar decomposition of the deformation tensor The following multiplicative decomposition of $\boldsymbol{F}$ is referred to as its polar decomposition and is given by

$$\boldsymbol{F} = \boldsymbol{R}\boldsymbol{U} = \boldsymbol{V}\boldsymbol{R}, \tag{2.5}$$

cf. Gurtin et al. (2010, p. 33). Here, $\boldsymbol{R} \in \mathit{Orth}^+$ describes the rigid-body rotation. The occurring stretches are given by the right and left stretch tensor $\boldsymbol{U}$ and $\boldsymbol{V}$, respectively. Both $\boldsymbol{U}$ and $\boldsymbol{V}$ are symmetric and positive definite, i.e., $\boldsymbol{U}, \boldsymbol{V} \in \mathit{Psym}$, and their eigenvalues coincide. The first relation of Eq. (2.5) is referred to as right polar decomposition and the second one as left polar decomposition. Using Eq. (2.5),

$$\boldsymbol{F}^{\mathsf{T}}\boldsymbol{F} = \boldsymbol{U}^{\mathsf{T}}\boldsymbol{R}^{\mathsf{T}}\boldsymbol{R}\boldsymbol{U} = \boldsymbol{U}^{\mathsf{T}}\boldsymbol{U} =: \boldsymbol{C}, \tag{2.6}$$

$$\boldsymbol{F}\boldsymbol{F}^{\mathsf{T}} = \boldsymbol{V}\boldsymbol{R}\boldsymbol{R}^{\mathsf{T}}\boldsymbol{V}^{\mathsf{T}} = \boldsymbol{V}\boldsymbol{V}^{\mathsf{T}} =: \boldsymbol{B} \tag{2.7}$$

holds true. The two tensors $\boldsymbol{C}$ and $\boldsymbol{B} \in Psym$ have the same eigenvalues and are referred to as the right and left Cauchy-Green tensor, cf. Bertram (2005). Consequently, for the eigenvalues μ_α and λ_α of $\boldsymbol{U}$ and $\boldsymbol{C}$, respectively, the following relationship applies

$$\lambda_\alpha = \mu_\alpha^2. \tag{2.8}$$

Deformation measures The eigenvalues of $\boldsymbol{U}$ are also denoted as principle stretches, cf. Marsden and Hughes (1994, p. 4). They are commonly used to define generalized strain measures as proposed by Hill (1968). A generalized strain measure of the Seth-type can be formulated as

$$\boldsymbol{E}^{\text{gen}} = \sum_{\alpha=1}^{3} f(\mu_\alpha)\boldsymbol{\beta}_\alpha \otimes \boldsymbol{\beta}_\alpha, \quad f(\mu_\alpha) = \begin{cases} \frac{\mu_\alpha^m - 1}{m}, & m \neq 0 \\ \\ \log \mu_\alpha, & m = 0 \end{cases} \tag{2.9}$$

cf. Bertram (2005, p. 114) and Hill (1968, Eq. (1)). Here, $\boldsymbol{\beta}_\alpha$ denotes an eigenvector of $\boldsymbol{U}$. The function $f(\mu_\alpha)$ is twice differentiable and fulfills the criteria

$$f' > 0 \; \forall \; \mu_\alpha, \quad f(1) = 0, \quad f(1)' = 1. \tag{2.10}$$

Specific choices of m lead to well known strain measures. Thus, Green's strain tensor is obtained for $m = 2$. Regarding small deformations, all strain tensors reduce to the infinitesimal strain tensor ε, reading

$$\varepsilon = \frac{1}{2}\left(\text{grad}\,(\boldsymbol{u}) + \text{grad}\,(\boldsymbol{u})^{\mathsf{T}}\right). \tag{2.11}$$

This property is independent of the choice of m. In case of small deformations, no difference is made between $\text{grad}\,(\cdot)$ and $\text{Grad}\,(\cdot)$.

Energy balance of a classical continuum Regarding the current configuration of a material volume, the balance of total energy of a classical continuum is given by

$$\frac{\mathrm{d}}{\mathrm{d}t}\int_{\mathcal{V}_t}\rho\left(e+\frac{1}{2}\boldsymbol{v}\cdot\boldsymbol{v}\right)\,\mathrm{d}v=\int_{\mathcal{V}_t}\rho\,(\boldsymbol{b}\cdot\boldsymbol{v}+r)\,\mathrm{d}v+\int_{\partial\mathcal{V}_t}\boldsymbol{t}\cdot\boldsymbol{v}+h\,\mathrm{d}a, \tag{2.12}$$

cf. Marsden and Hughes (1994). The volume of the continuum is referred to as $\mathcal{V}_t$ and its boundary as $\partial\mathcal{V}_t$. Equation (2.12) also holds true for a material volume that contains a singular surface. However, singular surfaces are not considered throughout this chapter. Here, e denotes the mass specific internal energy and ρ the mass density. The spatial velocity field of the body is denoted by $\boldsymbol{v}$. Mechanical power is expended by the body and traction forces $\boldsymbol{b}$ and $\boldsymbol{t}$, respectively. The thermal contribution is given by the heat supply r and the heat flux h.

Invariance consideration of the energy balance Given are two independent Euclidean vector spaces $\mathcal{W}$ and $\mathcal{W}'$. While Euclidean vector spaces are isomorphic in general, $\mathcal{W}$ and $\mathcal{W}'$ are distinguished, here. Each vector space is associated with an observer. The relation between quantities described by the corresponding observer is given by the Euclidean transformation

$$\boldsymbol{x}'(t)=\boldsymbol{Q}(t)\boldsymbol{x}(t)+\boldsymbol{c}'(t). \tag{2.13}$$

In this context, $\boldsymbol{Q}(t)$ describes a time-dependent isometry between the two vector spaces. Consequently, $\boldsymbol{Q}$ is invertible and $\det(\boldsymbol{Q})=1$ holds true. The origins of both vector spaces are related to each other by the time-dependent vector $\boldsymbol{c}'(t)$. Regarding $\mathcal{W}$ and $\mathcal{W}'$, $\boldsymbol{x}\in\mathcal{W}$, $\boldsymbol{x}'\in\mathcal{W}'$, $\boldsymbol{c}'\in\mathcal{W}'$ and $\boldsymbol{Q}:\mathcal{W}\to\mathcal{W}'$ holds true. Thus, the isometry $\boldsymbol{Q}$ is given by $\boldsymbol{Q}=Q_{ij}\boldsymbol{e}'_i\otimes\boldsymbol{e}_j$ with $\boldsymbol{e}'_i\in\mathcal{W}'$ and $\boldsymbol{e}_j\in\mathcal{W}$, cf. Krawietz (2015) for more details. Both observers consider the same physical process in their respective vector space. This motivates to assume the invariance of the energy balance with respect to a change of observer which yields the existence of the Cauchy stress tensor $\boldsymbol{\sigma}$ and the heat flux vector $\boldsymbol{q}$, reading

$$\boldsymbol{t} = \boldsymbol{\sigma}\boldsymbol{n}, \quad \text{and} \quad h = -\boldsymbol{q} \cdot \boldsymbol{n}, \tag{2.14}$$

cf. Šilhavý (1997). In this context, the surface normal vector is denoted as $\boldsymbol{n}$. Moreover, the balance of mass, linear and angular momentum, and the balance of internal energy are obtained

$$\begin{aligned} \dot{\rho} + \rho \operatorname{div}(\boldsymbol{v}) = 0, \quad \rho(\boldsymbol{a} - \boldsymbol{b}) - \operatorname{div}(\boldsymbol{\sigma}) = \boldsymbol{0}, \quad \boldsymbol{\sigma} = \boldsymbol{\sigma}^{\mathsf{T}}, \\ \rho\dot{e} - \rho r - \boldsymbol{\sigma} \cdot \boldsymbol{D} + \operatorname{div}(\boldsymbol{q}) = 0. \end{aligned} \tag{2.15}$$

The material time derivative is denoted by $\dot{(\;)}$. The acceleration is abbreviated as $\boldsymbol{a} = \dot{\boldsymbol{v}}$ and the symmetric part of the velocity gradient is denoted as

$$\boldsymbol{D} = \operatorname{sym}(\operatorname{grad}(\boldsymbol{v})). \tag{2.16}$$

A continuum satisfying the balance equations according to Eq. (2.12) and Eq. (2.15) is referred to as classical Boltzmann continuum or Cauchy continuum in the literature.

Entropy balance Regarding the current configuration of a material volume, the standard form of the entropy balance is given by

$$\frac{\mathrm{d}}{\mathrm{d}t} \int_{\mathcal{V}_t} \rho\eta \,\mathrm{d}v = -\int_{\partial\mathcal{V}_t} \boldsymbol{\phi}^{\eta} \cdot \boldsymbol{n} \,\mathrm{d}a + \int_{\mathcal{V}_t} \rho\, p^{\eta} + s^{\eta} \,\mathrm{d}v, \tag{2.17}$$

cf. Müller (1985). Here, η denotes the mass specific entropy considering the bulk material, $\boldsymbol{\phi}^{\eta}$ the entropy flux across the boundary $\partial\mathcal{V}_t$, p^{η} the mass specific entropy production, and s^{η} the entropy supply.

Clausius-Duhem inequality As common in classical thermodynamics, cf. Coleman and Noll (1963), the entropy flux is assumed to be given by $\boldsymbol{\phi}^{\eta} = \boldsymbol{q}/\theta$, and the entropy supply is assumed to be $s^{\eta} = \rho r/\theta$. The bulk dissipation is defined as $\delta := p^{\eta}\theta$. Accounting for the previous assumptions as well as Reynold's transport theorem and the divergence theorem, localization of Eq. (4.24) yields

$$\rho\delta = \rho\theta\dot{\eta} - \rho r + \theta \operatorname{div}\left(\frac{\boldsymbol{q}}{\theta}\right) \tag{2.18}$$

Moreover, the relation between the specific free energy ψ, the specific internal energy e and the specific entropy η is given as $\psi = e - \theta\eta$, resulting from the Legendre transformation, cf. Beegle et al. (1974). The second law of thermodynamics states that the dissipation is always non-negative. This yields the dissipation inequality based on Eq. (2.18), reading

$$\rho\delta = \rho\dot{e} - \rho\dot{\psi} - \rho\dot{\theta}\eta - \rho r + \operatorname{div}(\boldsymbol{q}) - \frac{1}{\theta}\boldsymbol{q}\cdot\boldsymbol{g} \geq 0, \tag{2.19}$$

with $\boldsymbol{g} = \operatorname{grad}(\theta)$. In this form, the dissipation inequality according to Eq. (2.19) is also often referred to as Clausius-Duhem inequality, cf. Coleman and Gurtin (1967).

Thermodynamically consistent material modeling The application of the Coleman-Noll procedure to the Clausius-Duhem inequality yields constitutive equations such as the potential relation for the stress tensor, cf. Coleman and Noll (1963); Coleman and Gurtin (1967). Due to its simplicity, this method is applied to an extended continuum in a subsequent chapter of the work at hand. However, there exist more elaborated procedures to obain thermodynamically consistent, constitutive restrictions. Among these is, e.g., the Müller-Liu procedure, cf. Liu (1972) and Müller (1985). Moreover, the so-called *GENERIC*-framework (general equation for non-equilibrium reversible-irreversible coupling) establishes a powerful method for the thermodynamical constitutive modeling, cf. Grmela and Öttinger (1997); Öttinger and Grmela (1997). Hütter and Svendsen (2011; 2013) formulate various material models in the context of *GENERIC* and discuss corresponding examples, respectively.

Principle of virtual power Mutliplication of the balance of linear momentum with a test function $\boldsymbol{f}$, integration over $\mathcal{V}_t$ and application of the divergence theorem yieds the corresponding weak form as

$$\int_{\mathcal{V}_t} \rho\,(\boldsymbol{a}-\boldsymbol{b})\cdot\boldsymbol{f}+\boldsymbol{\sigma}\cdot\operatorname{grad}(\boldsymbol{f})\,\mathrm{d}v-\int_{\partial\mathcal{V}_t}\boldsymbol{t}\cdot\boldsymbol{f}\,\mathrm{d}a=0. \tag{2.20}$$

Choosing the test function to be the variation of the spatial velocity, i.e., $\boldsymbol{f}=\delta\boldsymbol{v}$, yields the principle of virtual power (PoVP)

$$\int_{\mathcal{V}_t} \rho\,(\boldsymbol{a}-\boldsymbol{b})\cdot\delta\boldsymbol{v}+\boldsymbol{\sigma}\cdot\operatorname{grad}(\delta\boldsymbol{v})\,\mathrm{d}v-\int_{\partial\mathcal{V}_t}\boldsymbol{t}\cdot\delta\boldsymbol{v}\,\mathrm{d}a=0. \tag{2.21}$$

Similar to the balance laws stated in Eq. (2.12) and Eq. (2.15), the PoVP according to Eq. (2.21) is not restricted to a specific material behavior.

Variational principle Assuming a specific, material behavior as well as conservative body and traction forces, the PoVP according to Eq. (4.40) represents the stationary condition of a variational formulation, cf. Lanczos (1949) and Auffray et al. (2015, p. 379). Thus, the principle of least action serves as starting point for the continuum mechanical considerations. In both continuum mechanics and classical mechanics, this approach is referred to as analytical mechanics, cf. Auffray et al. (2015); Willner (2003), and Landau and Lifshitz (1969), respectively. In the context of nondissipative processes, Hamilton's principle of least action is widely used, cf. Planck (1960, p. 81), reading

$$\delta\mathcal{H}=0,\quad \mathcal{H}=\int_{t_0}^{t_1}\mathcal{L}\,\mathrm{d}t,\quad \mathcal{L}=\mathcal{E}_{\mathrm{kin}}-\Pi_{\mathrm{tot}}. \tag{2.22}$$

Here, $\mathcal{L}$ is referred to as Lagrange density and $\mathcal{H}$ as action integral. The kinetic energy is denoted as $\mathcal{E}_{\mathrm{kin}}$ and the potential energy as Π_{tot}. Regarding a classical continuum, the kinetic energy with respect to the current configuration reads

$$\mathcal{E}_{\mathrm{kin}} = \int_{\mathcal{V}_t} \frac{\rho}{2} \boldsymbol{v} \cdot \boldsymbol{v} \, \mathrm{d}v. \tag{2.23}$$

The potential energy Π_{tot} can be written as

$$\Pi_{\mathrm{tot}} = \int_{\mathcal{V}_t} \Phi_{\mathrm{V}} + W \, \mathrm{d}v + \int_{\partial\mathcal{V}_t} \Phi_{\mathrm{A}} \, \mathrm{d}a. \tag{2.24}$$

In this context, Φ_{V} and Φ_{A} denote the potentials of the conservative body and surface forces. Moreover, the stored energy is represented by W. Since conservative body and traction forces are considered, the following relations hold true

$$\delta\Phi_{\mathrm{V}} = \rho \boldsymbol{b} \cdot \delta\boldsymbol{u}, \quad \delta\Phi_{\mathrm{A}} = \boldsymbol{t} \cdot \delta\boldsymbol{u}. \tag{2.25}$$

Regarding a specific material, the stored free energy W can be given explicitly. In case of a hyperelastic material behavior W and δW read

$$W = \frac{1}{2} \boldsymbol{\varepsilon} \cdot (\mathbb{C}[\boldsymbol{\varepsilon}]), \quad \delta W = \boldsymbol{\sigma} \cdot \delta\boldsymbol{\varepsilon}, \quad \delta\boldsymbol{\varepsilon} = \mathrm{sym}(\delta\boldsymbol{u}). \tag{2.26}$$

The symmetry of the Cauchy stress $\boldsymbol{\sigma}$ is presumed from the outset. To account for dissipative processes, Hamilton's principle has to be extended by an additional potential, cf. Maugin (1980, p. 64) as well as Sedov (1968b;a). This so called dissipation potential represents a continuum mechanical equivalent of the Rayleigh potential which is commonly used in the context of discrete dynamics, cf. Ganghoffer (2007, p. 193). In contrast to the balance equations according to Eq. (2.12) and Eq. (2.15), and the PoVP according to Eq. (2.21), the material behavior, cf. Eq. $(2.26)_1$, has to be specified in order to state the considered Lagrange density used for a variational principle. Thus, a variational principle always accounts for constitutive assumptions, defining the material behavior, cf. Willner (2003, p. 172).

2.1.2 Generalized continua

The concept of a generalized continuum Any continuum model that is based on other assumptions than the classical continuum is denoted as an extended or generalized continuum, cf. Maugin (2017, p. 15). Generalized continua can be subdivided into local and nonlocal continuum models, cf. Forest (2019, p. 500). Regarding a local continuum model, it is possible to define the field equations point-wise. Contrarily, only integral formulations are available for nonlocal continuum models. The theory of nonlocal continua is described extensively by Eringen (2002). Thermodynamical considerations of nonlocal continua are discussed by Edelen and Laws (1971). Applications of this class of continua can be found, e.g., in the context of dislocations, cf. Eringen (1977a;b). Subsequently, only generalized continua of local type are considered. In this context, a material point of an extended continuum exhibits more than three translational DOFs. One of the first extended continuum models is the so called Cosserat continuum, cf. Cosserat and Cosserat (1909). It allows for the orientation of a material point. Thus, each material point is supplemented by three rotational DOFs in addition. The micromorphic continuum according to Eringen and Suhubi (1964) can be considered as a direct generalization of the Cosserat continuum. It treats each material point as a micro continuum. Consequently, a microdeformation tensor associated with each material point is introduced in Eringen and Suhubi (1964). Restricting the microdeformation to pure rotations yields the Cosserat continuum as a special case of the micromorphic continuum. Furthermore, prohibiting the rotation of a material point yields the classical Cauchy continuum. A detailed comparison of these continua, relations between them, as well as examples of application are given in Eringen (1999, p. 13). In Fig. 2.1 the Cauchy, Cosserat and micromorphic continuum are sketched. The Cosserat and micromorphic continuum account for extensions of kinematic nature. Moreover, it is possible to introduce more generic extensions that are not associated with an underlying kinematic,

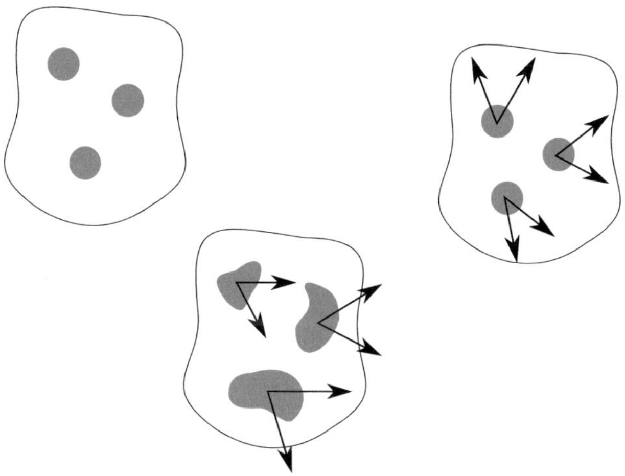

Figure 2.1: The upper left sketch illustrates the Cauchy continuum. A Cosserat continuum is depicted in the upper right sketch. The attached direction vectors illustrate the additional rotational DOFs. The micromorphic continuum is depicted in the lower sketch. The shape of the material points deviaties from a sphere, illustrating the microdeformation of a material point.

but significantly contribute to the mechanical behavior, cf. Svendsen (2004). Examples for generic DOFs are order or damage parameters, cf. Svendsen (1999), as well as the dislocation density, cf. Svendsen (2002). Subsequently, extensions of various kind are considered.

Microstructure function The kinematics of an additionally considered DOF is given by means of a microstructure function. The spatial velocity of, e.g., an additional scalar-valued DOF p is given by the material time derivative of the microstructure function $\tilde{\varphi}_t$, reading

$$\tilde{v}(\boldsymbol{x},t) = \left(\left. \frac{\partial \tilde{\varphi}_t(\boldsymbol{X})}{\partial t} \right|_{\boldsymbol{X}=\text{const.}} \right) \circ {\boldsymbol{\varphi}_t}^{-1}, \quad p = \tilde{\varphi}_t(\boldsymbol{X}). \tag{2.27}$$

Supplementary contributions that are power conjugated to $\tilde{v}(\boldsymbol{x},t)$ are accounted for by various approaches to generalized continua. The subsequent sections consider the invariance and covariance of an extended energy balance, an extended PoVP, and an extended variational approach. In this context, approaches from the literature are classified and the idea of the corresponding approach is roughly recited. The focus is on additional field equations or balance laws with respect to regular points. Thus, singular surfaces are not considered throughout this chapter. The tensorial order of the additional degree of freedom considered differs from theory to theory.

2.2 Methods Based on an Extended Energy Balance

2.2.1 Invariance of energy balance

The notion of invariance In literature, invariance considerations of the energy balance are formulated concerning a change of observer or a superimposed rigid body motion, respectively. As far as the derivation of balance equations is concerned, it makes no difference whether the invariance is assumed with regard to a superimposed rigid-body motion or a change of observer. However, it is essential to distinguish between these two methods when considering material behavior, cf. Svendsen (2001a, footnote 2), and Svendsen and Bertram (1999). Subsequently, only balance equations are discussed. Thus, the notion 'invariance of the energy balance' is used for both approaches. In general, an extended energy balance is considered with respect to the current configuration as

$$\epsilon = \frac{\mathrm{d}}{\mathrm{d}t}\int_{\mathcal{V}_t} \rho\left(e + \frac{1}{2}\boldsymbol{v}\cdot\boldsymbol{v} + \kappa\right)\,\mathrm{d}v - \int_{\mathcal{V}_t} \rho\,(\boldsymbol{b}\cdot\boldsymbol{v} + \beta + r)\,\mathrm{d}v - \int_{\partial\mathcal{V}_t} \boldsymbol{t}\cdot\boldsymbol{v} + s + h\,\mathrm{d}a = 0. \qquad (2.28)$$

The additional contributions κ, β and s differ for each considered approach. In most approaches, the extensions are related to additional kinematical

DOFs of various kinds. However, it is also possible to consider more general DOFs that are not associated to kinematics, cf., e.g., Dunn and Serrin (1986). Thermal contributions are often considered unaltered, compared to a classical continuum. Assuming invariance of Eq. (2.28) with respect to a change of observer, cf. Eq. (2.13), yields the following.

- No additional momentum-like balance equations are obtained, cf. Yavari and Marsden (2009), Prahs and Böhlke (2019b).
- The balance of internal energy deviates from its standard form, taking into account supplementary terms associated with the additional DOFs, cf. Maugin (1980, p. 63), and Yavari and Marsden (2009, p. 10).
- If the additional DOF is scalar-valued, the balance of angular momentum retains its standard form according to Eq. (2.15) and the stress tensor is symmetric.
- If the additional DOF is a vector or a tensor of higher order, the balance of angular momentum is affected as well by the additional contributions. In this case, the stress tensor is no longer necessarily symmetric.

The invariance properties of an extended energy balance is addressed in detail by Prahs and Böhlke (2019b), first considering a scalar-valued DOF and then a vectorial one. The idea to obtain conservation laws by means of an invariance consideration of the specific total energy is closely related to Noether's theorem, cf. the reprint Noether (1971).

Fundamental assumptions

- The primal quantity is the rate of energy which is balanced by power contributions.
- The power contributions are of mechanical and thermal nature.
- Mechanical power is expended by body and surface forces.
- Surface forces are assumed to depend on the normal vector of the surface.
- The lemma of Cauchy is fundamental to this theory, proving the existence of a stress tensor associated with the surface force. Some theories

directly give the surface power in terms of stress tensors and a heat flux vector, thereby implicitly taking into account the result of Cauchy's lemma. Nevertheless, these theories can also be formulated in terms of the corresponding surface forces, instead.

- Here, an Euclidean ambient space is considered, cf. Yavari and Marsden (2009, p. 1).
- The microstructure manifold is identified with the tangent space of the Euclidean space, cf. Yavari and Marsden (2009, p. 1).
- The material behavior is not specified from the outset.
- To consider a specific material behavior, the balance of entropy has to be introduced and exploited, in addition.

Micromorphic medium according to Eringen The micromorphic continuum according to Eringen and Suhubi (1964) consists of material points admitting a deformation by analogy to the common deformation of the material body. It is assumed, that a material point is described by its center of mass $\boldsymbol{X}$ and a corresponding direction vector Ξ. Both, $\boldsymbol{X}$ and Ξ are given regarding an arbitrary reference placement. In general, it is possible to specify a material point in more detail by several direction vectors Ξ_α with $\alpha = 1, \ldots, N$, leading to a micromorphic continuum of grade N, cf. Eringen (1970). However, the special case of a single direction vector is extensively discussed in the literature, cf. Eringen (1999), and considered in the following. The corresponding framework is denoted as micromorphic continuum of grade one, or birefly as micromorphic continuum. Three DOFs are associated with the translational motion of the centroids of the material points, referred to as macromotion or briefly motion $\boldsymbol{x} = \hat{\boldsymbol{x}}(\boldsymbol{X}, t)$. Moreover, nine DOFs are related to the (affine) deformation of each material point, denoted as micromotion $\boldsymbol{\xi} = \hat{\boldsymbol{\xi}}(\boldsymbol{X}, \Xi, t)$. The micromotion is given by the linear mapping $\boldsymbol{\xi} = \chi(\boldsymbol{X}, t)\Xi$, where χ denotes the microdeformation tensor, cf. Eringen (1999, p. 5). Consequently, the micromotion represents the deformation of the direction vector fixed at each material point. Thus, each material point admits twelve DOFs in contrast to three DOFs for a classical continuum, cf. Eringen (1992). The material time derivative of $\boldsymbol{\xi}$ is given by $\dot{\boldsymbol{\xi}} = \boldsymbol{\nu}_{\mathrm{E}}\boldsymbol{\xi}$, cf. Eringen (1992, Eq. (2.6)). The second-order tensor $\boldsymbol{\nu}_{\mathrm{E}}$ is referred to as gyration tensor. Related to the additional DOFs, the extensions to the energy balance are given as

$$\kappa = \frac{1}{2} (\boldsymbol{\nu}_{\mathrm{E}} \boldsymbol{i}_{\mathrm{E}}) \cdot \boldsymbol{\nu}_{\mathrm{E}}, \quad \beta = \boldsymbol{B}_{\mathrm{E}} \cdot \boldsymbol{\nu}_{\mathrm{E}}, \quad s = (\boldsymbol{m}_{\mathrm{E}}[\boldsymbol{\nu}_{\mathrm{E}}]) \cdot \boldsymbol{n}, \tag{2.29}$$

cf. Eringen (1992, Eq. (2.12), Eq. (3.1)). Here, the second-order tensors $\boldsymbol{i}_{\mathrm{E}}$ and $\boldsymbol{B}_{\mathrm{E}}$ denote the micro-inertia and the body moment density, respectively. The micro-inertia is associated with the additional DOFs, cf. Eringen (1992, Eq. (2.10)). It is formulated as a volumetric moment of second order, cf. Eringen (1999, Eq. 1.10.3). The stress moment tensor $\boldsymbol{m}_{\mathrm{E}}$ is a third order tensor. A relation between the micro-inertia of the reference and the current configuration is established by means of $\boldsymbol{\chi}$. The material time derivative of this relation is referred to as conservation of micro-inertia, cf. Eringen (1992, Eq. (2.11)). Note that, the conservation of micro-inertia constitutes rather an identity than a balance or conservation law. Thus, the designation as a conserved quantity is misleading. While the framework was originally introduced by Eringen and Suhubi (1964) and Suhubi and Eringen (1964), the notion of conservation of micro-inertia was first given in Eringen (1964). The conservation of mass of a micro-continuum is an axiom of the theory, cf. Eringen (1999, p. 31), Eringen and Suhubi (1964, p. 191). This yields the conservation of mass in its standard form as stated in Eq. (2.15). Invariance of the extended energy balance under superimposed rigid body motions, as given by Eq. (2.13), and a transformation of the form

$$\boldsymbol{\chi}' = \boldsymbol{Q}(t)\boldsymbol{\chi} \tag{2.30}$$

is stated, cf. Eringen (1999, Eq. 1.9.1, Eq. 1.9.6). Here, the orthogonal tensor $\boldsymbol{Q}(t)$ describes the rigid body rotation. Evidently, a time-dependent, rigid body rotation and translation of x, but only a rotation of χ is considered. The modifed balance of angular momentum is obtained by fulfilling the requirement of a vanishing scalar product of a second-order tensor and a skew symmetric tensor. This provides both possibilities that either the second-order tensor is symmetric or identical to zero. Eringen (1992) uses the more restrictive option and, thus, the vanishing second-order tensor constitutes the modified balance of angular momentum. A statement concerning the symmetry of the stress tensors is not provided. The corresponding modified balances of angular momentum and internal energy are given by Eringen (1992, Eq. (3.17), Eq. (3.19))[2]. While a brief summary and revision of the framework is given by Eringen (1992), the framework is elaborated to full extent in Eringen (1999). However, the notation in Eringen (1992) and Eringen (1999) differs slightly, cf. Eringen (1999, footnote 8 at p. 33). The subsequently presented microstretch and micropolar continua are special cases of the micromorphic continuum. They are based on restrictions concerning the introduced direction vectors and, thereby, the microdeformation χ.

Microstretch continuum according to Eringen Preventing a shearing of the direction vector of a micromorphic continuum, a so-called microstretch continuum ist obtained. In this context, the following decompositions of $\boldsymbol{B}_\mathrm{E}$ and $\boldsymbol{m}_\mathrm{E}$ are introduced

$$\boldsymbol{m}_\mathrm{E} = \frac{1}{3}\left(\boldsymbol{I} \otimes \boldsymbol{m}^{\langle 1 \rangle}\right)^{\mathsf{T}_\mathsf{H}} + \frac{1}{2}\left(\epsilon[\tilde{\boldsymbol{m}}^\mathsf{T}]\right)^{\mathsf{T}_\mathsf{H}}, \quad \boldsymbol{B}_\mathrm{E} = \frac{1}{3} b \boldsymbol{I} - \frac{1}{2}\epsilon[\tilde{\boldsymbol{B}}], \qquad (2.31)$$

cf. Eringen (1999, Eq. (2.1.8)). Here, $\boldsymbol{m}^{\langle 1 \rangle}$ is a vector, $\tilde{\boldsymbol{m}}$ a second-order tensor, and ϵ represents the permutation symbol. The identity of second order is given by $\boldsymbol{I}$. Moreover, the body moment density is decomposed by means of the scalar b and the vector $\tilde{\boldsymbol{B}}$. In addition, the gyration tensor $\boldsymbol{\nu}_\mathrm{E}$

[2] cf. also (Eringen, 1999, Eq. 2.2.32, Eq. 2.2.33)

and the micro-inertia tensor $\boldsymbol{i}_{\mathrm{E}}$ are decomposed as well according to

$$\boldsymbol{\nu}_{\mathrm{E}} = \nu \boldsymbol{I} - \boldsymbol{\epsilon}[\tilde{\boldsymbol{\nu}}], \quad \boldsymbol{i}_{\mathrm{E}} = \frac{1}{2} j_0 \boldsymbol{I} - \boldsymbol{j}, \quad \boldsymbol{j} = i_0 \boldsymbol{I} - \boldsymbol{i}_{\mathrm{E}}, \tag{2.32}$$

with $j_0 = \mathrm{sp}(\boldsymbol{j})$ and $i_0 = j_0/2$, cf. Eringen (1999, Eq. (1.10.18), Eq. (1.10.20)). Here, ν, j_0 and i_0 are scalar-valued, $\tilde{\boldsymbol{\nu}}$ represents a vector and $\boldsymbol{j}$ a second-order tensor. Accounting for these decompositions, the additional contributions to the energy balance are given by

$$\begin{aligned} \kappa &= \frac{1}{4} j_0 \nu^2 + \frac{1}{2} \boldsymbol{j} \cdot (\tilde{\boldsymbol{\nu}} \otimes \tilde{\boldsymbol{\nu}}), \quad \beta = \tilde{\boldsymbol{B}} \cdot \tilde{\boldsymbol{\nu}} + b\nu, \\ s &= \left(\tilde{\boldsymbol{m}}^{\mathsf{T}} \boldsymbol{n}\right) \cdot \tilde{\boldsymbol{\nu}} + \nu \boldsymbol{n} \cdot \boldsymbol{m}^{\langle 1 \rangle}, \end{aligned} \tag{2.33}$$

cf. Eringen (1999, Eq. (1.10.21), Eq. (2.1.9)). In contrast to the micromorphic continuum, the microstretch continuum provides six additional DOFs at each material point. In line with the micromorphic continuum model, the extended balance of energy is stated invariant with respect to a change of observer described by the transformations according to Eq. (2.13) and Eq. (2.30). The corresponding modifed balances of angular momentum and internal energy are given by Eringen (1999, Eq. (2.2.41-a), Eq. (2.2.42-a)).

Micropolar continuum according to Eringen The micropolar continuum is first introduced by Kafadar and Eringen (1971) and extensively discussed in Eringen (1999). Further limitation of the microstretch continuum to solely rotations of the direction vector leads to a micropolar continuum. Thereby, the decompositions according to Eqs. (2.31) and (2.32) are further specified as follows

$$\boldsymbol{m}^{\langle 1 \rangle} = \boldsymbol{0}, \quad b = 0, \quad \nu = 0, \tag{2.34}$$

cf. Eringen (1999, pp. 33, 39). Thus, the supplementary contributions to the energy balance are given by

$$\kappa = \frac{1}{2}\boldsymbol{j} \cdot (\tilde{\boldsymbol{\nu}} \otimes \tilde{\boldsymbol{\nu}}), \quad \beta = \tilde{\boldsymbol{B}} \cdot \tilde{\boldsymbol{\nu}}, \quad s = \left(\tilde{\boldsymbol{m}}^{\mathsf{T}} \boldsymbol{n}\right) \cdot \tilde{\boldsymbol{\nu}}, \tag{2.35}$$

cf. Eringen (1999, Eq. (2.1.10)). The micropolar continuum provides three additional DOFs in contrast to a micromorphic continuum. The balance of energy is stated again invariant with respect to a change of observer, cf. Eq. (2.13) and Eq. (2.30). Thereby obtained balances of angular and internal energy are modified and given by Eringen (1999, Eq. (2.2.46-a), Eq. (2.2.47-a)). The notion micropolar continuum is closely related to the Cosserat continuum. In contrast to a micropolar continuum, the Cosserat continuum does not provide a micro-inertia nor a conservation law of micro-inertia, cf. Eringen (1999, p. xiii, p. 12 footnote 3). Finally, neglecting any direction vectors fixed at a material point yields a classical continuum, denoted as Cauchy continuum. Further correlations concerning the kinematics and the obtained balances of the corresponding continuum due to the mentioned restrictions, as well as an extensive overview of the micromorphic, microstretch, and miropolar framework are given in Eringen (1999).

Higher order velocity gradients according to Green and Rivlin (1964c)
In this framework, the balance of energy is extended by contributions accounting for velocity gradients of higher order. Power expended by the second up to the Ω-th gradient of the velocity is taken into account. This is similar to the series evolution of the specific energy in terms of velocity gradients. The corresponding generalized forces, power conjugate to the velocity gradients, are denoted as surface and body force multipoles. In Green and Rivlin (1964c), the extension κ to the kinetic energy is neglected. However, the kinetic energy also has to be extended by higher-order velocity gradients, in general. It is also possible to account for the corresponding inertia effects by modifying the body force multipoles.

The term β is associated to body force multipoles and s to surface force multipoles. Thus, the considered extensions to the balance of energy are

$$\kappa = 0, \quad \beta = \sum_{\alpha=1}^{\Omega} \tilde{\boldsymbol{B}}^{(\alpha)} \cdot \operatorname{grad}^{(\alpha)}(\boldsymbol{v}), \quad s = \sum_{\alpha=1}^{\Omega} \tilde{\boldsymbol{T}}^{(\alpha)} \cdot \operatorname{grad}^{(\alpha)}(\boldsymbol{v}). \quad (2.36)$$

Here, $\operatorname{grad}^{(\alpha)}(\cdot)$ denotes the α-th gradient. Consequently, the tensors $\tilde{\boldsymbol{B}}^{(\alpha)}$ and $\tilde{\boldsymbol{T}}^{(\alpha)}$, power-conjugate to $\operatorname{grad}^{(\alpha)}(\boldsymbol{v})$, are tensors of order $(\alpha + 1)$. The postulated energy balance according to Green and Rivlin (1964c, Eq. 6.1) already implicitly accounts for the conservation of mass according to Eq. (2.15). Thus, it is presumed from the outset. Invariance of the extended energy balance is stated with respect to a superimposed rigid body motion, given by Eq. (2.13), cf. Green and Rivlin (1964c, Eq. 3.1). Contrarily to other theories, only the transformation law for $\boldsymbol{x}$ is considered since the additional contributions to the energy balance are higher-order velocity gradients. The modified balances of angular momentum and internal energy are given by Green and Rivlin (1964c, Eq. (6.15), Eq. (6.16)). This framework serves as predecessor for subsequent theories of Green and Rivlin, that account for other DOFs than the velocity gradients of arbitrary order.

Multipolar continuum mechanics according to Green and Rivlin (1964a) The multipolar continuum can be seen as a generalization of Green and Rivlin (1964c), cf. Green and Rivlin (1964a, comment on p. 113). Moreover, it serves as basis for the frameworks according to Green (1965); Green et al. (1965). Contrarily to Green and Rivlin (1964c), multipolar displacement fields $\boldsymbol{\chi}^{(\alpha)}$ are introduced instead of higher-order velocity gradients. The multipolar displacement fields are not related to the usual displacement field. They are tensors of $(\alpha + 1)$th order. Moreover, the extensions to the kinetic energy are accounted for by the term β. Hence, the introduced extensions to the energy balance are given by

$$\kappa = 0, \quad \beta = \sum_{\alpha=1}^{\Omega} \left(\boldsymbol{B}^{(\alpha)} - \boldsymbol{i}^{(\alpha)} \dot{\boldsymbol{\nu}}^{(\alpha)} \right) \cdot \boldsymbol{\nu}^{(\alpha)}, \quad s = \sum_{\alpha=1}^{\Omega} \boldsymbol{T}^{(\alpha)} \cdot \boldsymbol{\nu}^{(\alpha)}. \quad (2.37)$$

Here, $\boldsymbol{B}^{(\alpha)}$ and $\boldsymbol{T}^{(\alpha)}$ are generalized so-called multipolar body and surface forces. Moreover, $\boldsymbol{i}^{(\alpha)}$ is the corresponding micro-inertia tensor, and $\boldsymbol{\nu}^{(\alpha)}$ the time derivative of $\boldsymbol{\chi}^{(\alpha)}$. The tensors $\boldsymbol{B}^{(\alpha)}$, $\boldsymbol{T}^{(\alpha)}$, $\boldsymbol{\nu}^{(\alpha)}$ and $\boldsymbol{i}^{(\alpha)}$ are tensors of order $(\alpha + 1)$. The extended energy balance is stated invariant under a superimposed rigid body motion, defined by Eq. (2.13). In this context, the mutlipolar displacement fields are assumed to transform objectively with respect to a rigid body rotation. Thus, for, e.g., $\alpha = 1$, it follows

$$\boldsymbol{\chi}'^{(1)} = \boldsymbol{Q}(t)\boldsymbol{\chi}^{(1)}\boldsymbol{Q}(t)^{\mathsf{T}}, \quad (2.38)$$

cf. Green and Rivlin (1964a, Eq. 4.6). The modified balances of angular momentum and internal energy are given by Green and Rivlin (1964a, Eq. (8.19), Eq. (8.21)). Here, the same definitions of additionally introduced body and surface forces are considered as in Truesdell and Toupin (1960, Sec. 232), cf. Green and Rivlin (1964a, remark on p. 114). Furthermore, conditions for the applicability of the mentioned framework of Truesdell and Toupin (1960), which is based on an extended virtual work formulation, are given. A comparison of the derivation and the framework of Green and Rivlin (1964a) with Eringen and Suhubi (1964) and Eringen (1964) is given in Green (1965).

Director theory according to Green et al. (1965) The director theory is a special case of the framework given by Green and Rivlin (1964a). It can be obtained by choosing $\Omega = 2$. The director $\boldsymbol{d}$ is defined by the relation $\boldsymbol{\chi}^{(2)} = \boldsymbol{d} \otimes \boldsymbol{I}$. A detailed discussion on this is given in Green et al.

(1965, p. 614). The extensions to the energy balance are given by

$$\kappa = 0, \quad \beta = \boldsymbol{B}^{(1)} \cdot \boldsymbol{\nu}^{(1)} + \tilde{\boldsymbol{b}} \cdot \boldsymbol{\omega}, \quad s = \boldsymbol{T}^{(1)} \cdot \boldsymbol{\nu}^{(1)} + \tilde{\boldsymbol{t}} \cdot \boldsymbol{\omega}. \tag{2.39}$$

While the director velocity $\boldsymbol{\omega}$ is a vector, the velocity $\boldsymbol{\nu}^{(1)}$ of the multipolar displacement is a second-order tensor. Following the notation of Green and Rivlin (1964c), $\boldsymbol{B}^{(1)}$ and $\tilde{\boldsymbol{b}}$ are denoted as mutlipolar body forces, while $\boldsymbol{T}^{(1)}$ and $\tilde{\boldsymbol{t}}$ are referred to as multipolar surface forces. The micro-inertia effects associated with $\boldsymbol{\nu}^{(1)}$ and $\boldsymbol{\omega}$ are accounted for by $\boldsymbol{B}^{(1)}$ and $\tilde{\boldsymbol{b}}$ implicitly. However, they are not given explicitly by Green et al. (1965). The invariance of the extended energy balance is stated concerning the superimposed rigid body motion given in Eq. (2.13). Moreover, the transformations for the director and the multipolar displacement field read

$$\boldsymbol{\chi}'^{(1)} = \boldsymbol{Q}(t)\boldsymbol{\chi}^{(1)}\boldsymbol{Q}(t)^{\mathsf{T}}, \quad \boldsymbol{d}' = \boldsymbol{Q}(t)\boldsymbol{d}, \tag{2.40}$$

cf. Green et al. (1965, Eq. (2.6), Eq. (2.7)). The modified balances of angular momentum and internal energy are given by Green et al. (1965, Eq. (4.21), Eq. (4.20)). If only the director $\boldsymbol{d}$ is considered as additional DOF and the mutlipolar displacement of first order is neglected, i.e., $\boldsymbol{\nu}^{(1)} = \boldsymbol{0}$, the theory reduces to that of Ericksen (1961), cf. Green et al. (1965, p. 618). In this case, the balances of angular momentum and internal energy are further modified and given by Green et al. (1965, Eq. (4.31), Eq. (4.30)).

Micro-materials according to Green (1965) This continuum also constitutes a special case of the multipolar continuum according to Green and Rivlin (1964a). It provides a comparison between the theory according to Eringen and Suhubi (1964); Eringen (1964) and the theories of Green, especially with respect to Green and Rivlin (1964a). To this end, only a mutlipolar displacement field of first order is considered as additional DOF. The multipolar displacement field is chosen according to the theory of Eringen and Suhubi (1964). Thus, the additional contributions to the energy balance are in line with Eq. (2.29). The extended energy balance is stated invariant with respect to a change of observer described by Eq. (2.13) and Eq. (2.30). The modified balances of angular momentum and internal energy are given in Green (1965, Eq. (3.16), Eq. (3.14)). Similar to the micromorphic continuum, the conservation of micro-inertia is obtained, cf. Green (1965, Eq. (3.19)) and Eringen (1992, Eq. (2.11)).

Affine degrees of freedom according to Capriz et al. (1982) In order to describe the deformation of the microstructure of a material point, a second-order tensor $\boldsymbol{G}$ is introduced in Capriz et al. (1982). This approach is quite similar to Eringen's micromorphic theory, i.e., the mapping $\boldsymbol{\chi}$ resembles $\boldsymbol{G}$. Contrarily to the micromorphic continuum, the framework of Capriz et al. (1982) is not associated to the deformation of a direction vector fixed at a material point. In fact, it is a generalization of the deformation of the microstructure. Thus, it accounts for, e.g., rigid granular materials and materials with voids. Based on the kinematic descriptor $\boldsymbol{G}$, the tensor $\boldsymbol{W}^{\mathsf{T}} = \dot{\boldsymbol{G}}\boldsymbol{G}^{-1}$ is introduced in order to formulate the extension of the kinetic and mechanical energy contributions. The considered extensions are given by

$$\kappa = \frac{1}{2}(\boldsymbol{i}\boldsymbol{W}) \cdot \boldsymbol{W}, \quad \beta = \boldsymbol{B} \cdot \boldsymbol{W}, \quad s = (\boldsymbol{m}\boldsymbol{n}) \cdot \boldsymbol{W}. \tag{2.41}$$

The micro-inertia $\boldsymbol{i}$ is a second-order, symmetric and positive semi-definite tensor. Moreover, $\boldsymbol{B}$ and $\boldsymbol{m}$ are tensors of second and third order, respectively. The surface normal is again denoted as $\boldsymbol{n}$. Invariance of the extended energy balance is stated with respect to a change of observer, described by Eq. (2.13). The microstructure tensor $\boldsymbol{G}$ is assumed to transform according to

$$\boldsymbol{G}' = \boldsymbol{Q}(t)\boldsymbol{G}. \tag{2.42}$$

Based on the proposed invariance properties, the modified balances of angular momentum and internal energy are given by Capriz et al. (1982,

Eq. (4.11), Eq. (4.17)). In addition, the conservation of micro-inertia is obtained in Capriz et al. (1982, Eq. (4.12))[3]. This is in contrast to Eringen's framework where conservation of mass and microinertia are not derived by invariance considerations, cf. Eringen (1999, p. 32).

Additional, generic scalar DOFs The extension of the energy balance does not require the introduction of additional kinematic quantities, such as a microdeformation or higher velocity gradients. Moreover, the additionally introduced DOFs can be of a more general nature. For instance, the movement of point effects or the density of geometrically necessary dislocations can be accounted for as additional DOFs, cf. Svendsen (2001a; 2002). Moreover, the additional DOFs can be related to order parameters of different phases regarding phase transition, or the volume fraction of grains in the context of granular media, cf. Svendsen (1999). Thus, an internal length scale is introduced associated with the considered additional, scalar valued DOFs. While the energy balance is extended by means of several scalar valued DOFs in Svendsen (1999; 2001a; 2002; 2004), only one additional DOF is considered, here, for brevity. Hence, the extensions to the energy balance are given as

$$\kappa = \frac{1}{2}\tilde{A}\tilde{v}^2, \quad \beta = \tilde{b}\tilde{v}, \quad s = \tilde{t}\tilde{v}, \tag{2.43}$$

cf. Prahs and Böhlke (2019b). The proposed energy balance is stated to be Euclidean frame-indifferent, i.e., invariant under the transformation described by Eq. (2.13). The scalar valued additional DOF as well as its rate $\tilde{v}$, and the conjugated forces $\tilde{b}$ and $\tilde{t}$, as well as the micro-inertia $\tilde{A}$ are invariant with respect to Eq. (2.13). Since the considered additional DOF is scalar-valued, the balance of angular momentum retains its standard form, cf. Prahs and Böhlke (2019b, Eq. (31)). However, the balance of internal energy is affected by the additional contributions as given by Prahs and Böhlke (2019b, Eq. (36)). In case of a scalar-valued DOF, a conservation

[3] In Capriz et al. (1982) the dot placed over $\imath$, denoting the material time derivative, is missing in Eq. 4.12. However, this issue is clarified by the corresponding, given footnote.

of micro-inertia cannot be obtained by invariance considerations. Based on the contribution power-conjugated to $\tilde{v}$, a nonlocal material behavior can be constitutively modeled regarding the dissipation inequality. In Svendsen (1999) this is done by means of the Müller-Liu procedure, cf. Müller (1985).

The special case of isothermal processes is extensively described by Svendsen (2011), regarding not only scalar-valued DOFs but also additional DOFs of higher tensorial order. Thereby, the energy balance considered differs from Eq. (2.28). It directly takes into account the constitutive assumptions used in the context of the Clausius-Duhem inequality.

The interstitial work hardening framework according to Dunn and Serrin (1986, Eq. 1.9) is obtained if the kinetic and volumetric extensions are neglected, i.e., $\kappa = 0$ and $\beta = 0$, and the surface contribution is simplified as $s = u$, with u as the interstitial working.

Formal extension of the energy balance An energy balance, differing from the energy balance given in Eq. (2.12) with respect to its form, is derived in the appendix of Green and Naghdi (1995a). This energy balance is extended in order to account for additional vector fields in Green and Naghdi (1995b). Consequently, its form also differs from the energy balance given in Eq. (2.28). In order to keep the extended continuum generic, the additionally introduced kinematic quantities are no further specified. However, connection is established to a Cosserat continuum with a single director. Regarding the invariance discussions of the previously recited frameworks, the additional DOF was affected only by the rotational transformation, if it is vector-valued or a tensor of higher order. Contrarily, an additional transformation is considered in Green and Naghdi (1995b). It is assumed, that the introduced director exhibits a translational transformation not related to the translational transformation of the classical displacement, cf. Green and Naghdi (1995b, Eq. (2.17)). Invariance of the extended energy balance with respect to the translational transformations is stated, cf. (Green and Naghdi, 1995b, Eq. (2.18)). In contrast to all previously presented frameworks, it is not possible to obtain a modified balance of angular momentum, cf. (Green and Naghdi, 1995b, p. 365). However, a modified balance of momentum, an additional balance of director momentum and conservation laws for micro-inertia coefficients are obtained, cf. (Green and Naghdi, 1995b, Eq. (2.29), Eq. (2.30), Eq. (2.31)). However, the arbitrariness of the considered, independent rigid body translations give rise to criticism, cf. (Yavari and Marsden, 2009, p. 8). A critical point is that the additional director is introduced on the same manifold as the displacements. The introduced rigid body translations seem to be artificially introduced in order to derive the additional balance equations. Furthermore, no physical nor mathematical motivation is presented with respect to these transformations.

Summary of the considered theories Regarding the previously outlined extended continua, an overview of the considered additional contributions is given in Tab. 2.1. Moreover, the corresponding references are listed.

Theory	κ	β	s	associated works
Micro-morphic	$\frac{1}{2}\left(\boldsymbol{\nu}_{\mathrm{E}}\boldsymbol{i}_{\mathrm{E}}\right)\cdot\boldsymbol{\nu}_{\mathrm{E}}$	$\boldsymbol{B}_{\mathrm{E}}\cdot\boldsymbol{\nu}_{\mathrm{E}}$	$(\boldsymbol{m}_{\mathrm{E}}[\boldsymbol{\nu}_{\mathrm{E}}])\cdot\boldsymbol{n}$	Eringen and Suhubi (1964)
Microstretch	$\frac{1}{4}j_0\nu^2+\frac{1}{2}\boldsymbol{j}\cdot(\tilde{\boldsymbol{\nu}}\otimes\tilde{\boldsymbol{\nu}})$	$\tilde{\boldsymbol{B}}\cdot\tilde{\boldsymbol{\nu}}+b\nu$	$\left(\tilde{\boldsymbol{m}}^{\mathsf{T}}\boldsymbol{n}\right)\cdot\tilde{\boldsymbol{\nu}}+\nu\boldsymbol{n}\cdot\boldsymbol{m}^{\langle 1\rangle}$	Eringen (1999)
Micropolar	$\frac{1}{2}\boldsymbol{j}\cdot(\tilde{\boldsymbol{\nu}}\otimes\tilde{\boldsymbol{\nu}})$	$\tilde{\boldsymbol{B}}\cdot\tilde{\boldsymbol{\nu}}$	$\left(\tilde{\boldsymbol{m}}^{\mathsf{T}}\boldsymbol{n}\right)\cdot\tilde{\boldsymbol{\nu}}$	Eringen (1999); Kafadar and Eringen (1971)
Higher gradient	0	$\sum_{\alpha=1}^{\Omega}\tilde{\boldsymbol{B}}^{(\alpha)}\cdot\mathrm{grad}^{(\alpha)}(\boldsymbol{v})$	$\sum_{\alpha=1}^{\Omega}\tilde{\boldsymbol{T}}^{(\alpha)}\cdot\mathrm{grad}^{(\alpha)}(\boldsymbol{v})$	Green and Rivlin (1964c)
Multipolar	0	$\sum_{\alpha=1}^{\Omega}\left(\boldsymbol{B}^{(\alpha)}-\boldsymbol{i}^{(\alpha)}\dot{\boldsymbol{\nu}}^{(\alpha)}\right)\cdot\boldsymbol{\nu}^{(\alpha)}$	$\sum_{\alpha=1}^{\Omega}\boldsymbol{T}^{(\alpha)}\cdot\boldsymbol{\nu}^{(\alpha)}$	Green and Rivlin (1964a)
Director	0	$\boldsymbol{B}^{(1)}\cdot\boldsymbol{\nu}^{(1)}+\tilde{\boldsymbol{b}}\cdot\boldsymbol{\omega}$	$\boldsymbol{T}^{(1)}\cdot\boldsymbol{\nu}^{(1)}+\tilde{\boldsymbol{t}}\cdot\boldsymbol{\omega}$	Green et al. (1965)
Affine DOFs	$\frac{1}{2}\left(\boldsymbol{i}\boldsymbol{W}\right)\cdot\boldsymbol{W}$	$\boldsymbol{B}\cdot\boldsymbol{W}$	$(\boldsymbol{m}\boldsymbol{n})\cdot\boldsymbol{W}$	Capriz et al. (1982)
Scalar DOFs	$\frac{1}{2}\tilde{A}\tilde{v}^2$	$\tilde{b}\tilde{v}$	$\tilde{t}\tilde{v}$	Prahs and Böhlke (2019b); Svendsen (1999)
Interstitial	0	0	u	Dunn and Serrin (1986)

Table 2.1: List of theories that rest on invariance considerations of an extended energy balance. The corresponding additional contributions to the energy balance are listed. Each theory is associated with a corresponding paper.

Limitations concerning the Euclidean space as ambient space Subsequently, the limitations of the previously described extended continua, except the formal extension according to Green and Naghdi (1995b), are given. This theory is deliberately omitted, as the transformations listed there lack any physical or mathematical justification.

- The consideration of additional DOFs only affects the balance of angular momentum and the balance of internal energy.
- However, no additional balance equation associated with the supplementary DOFs is obtained.
- This is caused by the fact that the considered ambient space is an Euclidean space instead of, e.g., a Riemannian manifold, cf. Yavari and Marsden (2009).
- The additional DOFs are defined on the tangent space of the current configuration of the deformed body, cf. Capriz et al. (1982, p. 81). Hence, they are also defined on an Euclidean space. Consequently, they are rotated by the same transformation as the macro fields, i.e., by $\boldsymbol{Q}(t)$. However, a rigid body translation is not considered in this context. Thus, the additional DOFs can only enter the conservation law which is associated with the invariance under rotational transformation, namely the balance of angular momentum. This is similar to the implications of Noether's theorem, cf., e.g., the reprint Noether (1971).
- If the additional DOFs are scalar valued, they are commonly assumed invariant under a change of observer, cf. Svendsen (1999). In this case, the balance of angular momentum remains in its standard form. Only the balance of internal energy is modified due to the additional contributions.
- Some extended continua provide a conservation law for the micro-inertia. Regarding a scalar-valued DOF, the conservation of micro-inertia cannot be shown by invariance considerations, cf. Prahs and Böhlke (2019b).

2.2.2 Covariance of energy balance

Motivation for the generalisation of the ambient space Classical continuum mechanics is based on the assumption, that a three dimensional body is embedded in the Euclidean space, cf., e.g., Noll (1958, p. 200). The deformation of the body and accompanied stresses are described with respect to an arbitrary, stress free reference configuration that is also embedded in the Euclidean space. This viewpoint holds true if a body can be regarded phenomenologically as a simply connected set. If the deformation of a body has to be described in more detail, its microstructure has to be accounted for. This incorporates, e.g., dislocations and defects that are already present and quantifiable in the reference configuration. For instance, it is possible to measure an initial dislocation density for a given specimen. Consequently, describing a body as a generic manifold, rather than a subset of Euclidean space, seems to be more adequate, cf. Yavari and Goriely (2012, p. 60) and Maugin (2017). Moreover, this topic is already addressed by Kondo (1955); Bilby et al. (1955). This entails a geometric treatment of continuum mechanics as proposed by, e.g., Marsden and Hughes (1994). Several approaches can be found in literature, cf. Panoskaltsis and Soldatos (2014); Yavari and Goriely (2012); Yavari and Marsden (2009). The framework of Yavari and Marsden (2009) is briefly summarized in the following.

Transition to Riemannian manifolds In this context, the reference configuration is defined as a so-called material manifold $\mathcal{B}$, whereas the current configuration is defined as spatial manifold $\mathcal{C}$, cf., e.g., Marsden and Hughes (1994). Moreover, the considered microstructure is assigned to a separate microstructure manifold $\mathcal{M}$. The elements of $\mathcal{B}, \mathcal{C}$ and $\mathcal{M}$ are denoted as $\boldsymbol{X}$, $\boldsymbol{x}$ and $\boldsymbol{p}$, respectively, cf. Yavari and Marsden (2009, p. 4). They are related by the mappings

$$\boldsymbol{x} = \varphi_t(\boldsymbol{X}), \quad \boldsymbol{p} = \tilde{\varphi}_t(\boldsymbol{X}), \quad \psi_t = \tilde{\varphi}_t \circ {\varphi_t}^{-1}, \tag{2.44}$$

cf. Yavari and Marsden (2009, pp. 4,5). A graphical illustration is given by Yavari and Marsden (2009, Fig. 2.1). Consequently, the energy balance is formulated with respect to these manifolds and, thus, more generic. Instead of the scalar products in Eq. (2.12), inner products $\langle\langle \cdot, \cdot \rangle\rangle_{g}$ are considered with respect to the metric $\boldsymbol{g}$ of the spatial manifold $\mathcal{C}$. A concise comparison between classical tensor algebra and tensor algebra on manifolds is given by Stumpf and Hoppe (1997). The common contributions of the energy balance are replaced as follows

$$\boldsymbol{v} \cdot \boldsymbol{v} \rightarrow \langle\langle \boldsymbol{v}, \boldsymbol{v} \rangle\rangle_{\boldsymbol{g}}, \quad \boldsymbol{b} \cdot \boldsymbol{v} \rightarrow \langle\langle \boldsymbol{b}, \boldsymbol{v} \rangle\rangle_{\boldsymbol{g}}, \quad \boldsymbol{t} \cdot \boldsymbol{v} \rightarrow \langle\langle \boldsymbol{t}, \boldsymbol{v} \rangle\rangle_{\boldsymbol{g}}. \tag{2.45}$$

The covariance of the energy balance of a classical Cauchy continuum is extensively discussed by Yavari et al. (2006) and Marsden and Hughes (1994, pp. 165-167). Naturally, the velocity field is an element of the tangent space $T_x\mathcal{C}$ of $\mathcal{C}$, cf. Yavari and Marsden (2009, Eq. (2.1)). Here, $\boldsymbol{b}$ and $\boldsymbol{t}$ are also introduced as elements of $T_x\mathcal{C}$, i.e., as vectors, cf. Marsden and Hughes (1994). It is also possible to introduce $\boldsymbol{b}$ and $\boldsymbol{t}$ as elements of the cotangent space $T_x^*\mathcal{C}$ of $\mathcal{C}$, i.e., as one-forms. This bears the advantage that the notion of power is defined without the consideration of a metric, cf. Kanso et al. (2007). However, this is omitted, here.

The notion of covariance Here, the covariance of the energy balance instead of its invariance concerning a change of observer is stated, cf. Marsden and Hughes (1994). The covariance of the energy balance connotes its invariance under arbitrary diffeomorphisms. It is assumed, that $\mathcal{C}$ and $\mathcal{M}$ are independent of each other, i.e., that the transformation of $\mathcal{C}$ does not affect $\mathcal{M}$, and vice versa. This is one of the approaches proposed by Yavari and Marsden (2009). Consequently, instead of the transformation according to Eq. (2.13), the push-forward of $\boldsymbol{x}$ and $\boldsymbol{p}$ by the diffeomorphisms $\boldsymbol{\xi}_t$ and $\boldsymbol{\eta}_t$ is considered, respectively, reading

$$\boldsymbol{x}' = \boldsymbol{\xi}_{t*}\boldsymbol{x}, \quad \boldsymbol{p}' = \boldsymbol{\eta}_{t*}\boldsymbol{p}. \tag{2.46}$$

Here, $\boldsymbol{\xi}_t$ is referred to as spatial and $\boldsymbol{\eta}_t$ as microstructure diffeomorphism. The subscript $*$ denotes the push-forward.

Fundamental assumptions Regarding the covariance of the energy balance, the primal assumptions are comparable to those listed in Section 2.2.1. However, the following points are significantly different:

- Here, the ambient space is a Riemannian manifold instead of an Euclidean space, cf. Yavari and Marsden (2009, p. 10).
- The additional DOFs are element of an additionally introduced Riemannian manifold $\mathcal{M}$ representing the microstructure, cf. Yavari and Marsden (2009, p. 4).
- The specific internal energy e depends not only on $\boldsymbol{x}$ and t but also on the metric tensors $\boldsymbol{g}$ and $\tilde{\boldsymbol{g}}$ of the manifolds $\mathcal{C}$ and $\mathcal{B}$, and on the additional DOF $\boldsymbol{p}$, cf. Yavari and Marsden (2009, Eq. (4.8)).
- A specific material behavior is taken into account implicitly. The transformation of the stress vector with respect to an arbitrary diffeomorphism is specified for a material class, cf. Marsden and Hughes (1994, pp. 163, 164). While it is clear regarding pure elasticity, an altered transformation is necessary, e.g., for plasticity.

Supplementary contributions associated with an additional vectorial DOF In general, additional DOFs are introduced as elements of $\mathcal{M}$. Following the generic framework of Yavari and Marsden (2009), there exists no restriction on the tensorial order of the considered additional DOFs. For brevity, one vectorial DOF $\boldsymbol{p}(\boldsymbol{X}, t)$ is introduced, which is referred to as director field. Its velocity is denoted as $\tilde{\boldsymbol{v}}$. Thus, the additional contributions to the energy balance are given by

$$\kappa = \tilde{A}\langle\langle \tilde{\boldsymbol{v}}, \tilde{\boldsymbol{v}} \rangle\rangle_{\tilde{\boldsymbol{g}}}, \quad \beta = \langle\langle \tilde{\boldsymbol{b}}, \tilde{\boldsymbol{v}} \rangle\rangle_{\tilde{\boldsymbol{g}}}, \quad s = \langle\langle \tilde{\boldsymbol{t}}, \tilde{\boldsymbol{v}} \rangle\rangle_{\tilde{\boldsymbol{g}}}. \tag{2.47}$$

While $\tilde{\boldsymbol{b}}$ is denoted as micro-body force, $\tilde{\boldsymbol{t}}$ is referred to as micro-traction force. In contrast to, e.g., the micromorphic continuum according to

Eringen (1964), the micro-inertia is given by the scalar $\tilde{A}$, here. The quantities $\tilde{\boldsymbol{v}}, \tilde{\boldsymbol{b}}$ and $\tilde{\boldsymbol{t}}$ are elements of the tangent space $T_p\mathcal{M}$, cf. Yavari and Marsden (2009, p. 4).

Implications of the stated covariance Covariance of the extended energy balance under the spatial diffeomorphism yields conservation of mass as well as the standard balances of linear and angular momentum, given by Eq. (2.15). It is explicitly emphasized by Marsden and Hughes (1994) that the stress vector transforms objectively, if the considered spatial diffeomorphism $\boldsymbol{\xi}_t$ describes a rigid deformation. A rigid deformation can always be expressed by an Euclidean transformation. This statement holds true irrespective of the underlying material behavior. However, regarding the assumed, generic diffeomorphism, an objective transformation of the stress vector is only postulated in the purely elastic case, cf. Marsden and Hughes (1994, p. 163, bottom line). This is in contrast to all previously discussed extended continua that are derived by means of the invariance of the energy balance under a change of observer or a superimposed rigid body motion. Since the internal energy is assumed to depend on the metric of $\mathcal{C}$, a constitutive equation is obtained, in addition. It is denoted as Doyle-Ericksen formula, cf. Marsden and Hughes (1994), and serves as potential relation for the Cauchy stress in the purely elastic case reading

$$\boldsymbol{\sigma} = 2\rho \frac{\partial e}{\partial \boldsymbol{g}}. \tag{2.48}$$

Further, covariance with respect to the microstructure diffeomorphism leads to the conservation of micro-inertia and as well as balances of linear and angular microstructure momentum. Thus, the additionally obtained balance equations are given by

$$\dot{\tilde{A}} = 0, \quad \rho\left(\tilde{A}\tilde{\boldsymbol{a}} - \tilde{\boldsymbol{b}}\right) - \operatorname{div}\left(\tilde{\boldsymbol{\sigma}}\right) = \boldsymbol{0}, \quad \left(\boldsymbol{F}_0\tilde{\boldsymbol{\sigma}}\right)^{\mathsf{T}} = \boldsymbol{F}_0\tilde{\boldsymbol{\sigma}}. \tag{2.49}$$

The conservation of micro-inertia is given by Eq. $(2.49)_1$. Contrarily to the theories presented in Section 2.2.1 an additional balance equation for the additional DOF is obtained. This balance, given by Eq. $(2.49)_2$, resembles the balance of linear momentum. Therefore, it is often referred to as micro-force balance. The symmetry properties of the micro-stress tensor $\tilde{\boldsymbol{\sigma}}$ is described by Eq. $(2.49)_3$. Here, $\boldsymbol{F}_0$ is the differential $T\psi_t$. Its relation to the differentials of $\boldsymbol{\varphi}_t$ and $\tilde{\varphi}_t$ is given by

$$\boldsymbol{F} = T\boldsymbol{\varphi}_t, \quad \tilde{\boldsymbol{F}} = T\tilde{\varphi}_t, \quad \boldsymbol{F}_0 = \tilde{\boldsymbol{F}}\boldsymbol{F}^{-1}. \tag{2.50}$$

As a consequence of the assumed independence of the considered manifolds, the balance equations on micro scale according to Eq. (2.49) are completely decoupled from the standard balance equations given by Eq. (2.15). A relation between the fields of the common and the microstructure balances can be established by means of constitutive equations, i.e., due to exploitation of the second law of thermodynamics. Another approach is to assume constraints between the manifolds $\mathcal{C}$ and $\mathcal{M}$ as discussed in Yavari and Marsden (2009). However, this exceeds the scope of the review at hand, and is not discussed.

Limitations of the covariance considerations

- The transformation of the stress vector depends on the considered material behavior. The summarized framework according to Yavari and Marsden (2009) is limited to elastic materials. The same holds true for the treatment of a classical continuum according to Marsden and Hughes (1994, pp. 163, 164). This limitation is regarded as its most critical point.
- The consideration of dissipative processes by means of this framework is quite involved. Nondissipative processes are accounted for by internal variables in Panoskaltsis et al. (2013, p. 2107). They state that for this approach the covariance of the energy balance cannot hold true, cf. Panoskaltsis et al. (2013, p. 2116). In order to circumvent this

issue, a dissipative process is described by means of micro-forces that act on additional DOFs instead by Panoskaltsis and Soldatos (2014). However, both approaches assume the transformation of the stress vector as used in the purely elastic case. This choice is not justified according to Marsden and Hughes (1994, p. 163).

- Following Truesdell and Toupin (1960), a clear separation between balance equations and constitutive equations has to be drawn. This arises from the demand that balance equations should be of generic nature, and, thus, valid for all materials. Hence, as stated by Truesdell and Toupin (1960, p. 529), constitutive laws cannot be obtained from balance equations. Contrarily, the covariance of the energy balance provides a potential relation for the Cauchy stress in terms of the Doyle-Ericksen formula, cf. Marsden and Hughes (1994, p. 167) and Yavari and Marsden (2009, Eq. (4.24)).

2.3 Extended Principle of Virtual Power

2.3.1 Fundamentals

Principle of virtual power or virtual work for extended continua As already mentioned in Section 2.1.1, variational principles such as the principle of virtual work (PoVW) or the principle of virtual power (PoVP), cf. Eq. (4.40), can be used axiomatically to introduce continuum mechanics, cf. Willner (2003, p. 172) and Willner (2003, references at p. 175). Evidently, it can be used as a starting point for the formulation of extended continua.

In this context, an extended variational principle is given by

$$\delta\mathcal{P} = \delta\mathcal{P}_{\mathrm{a}} + \delta\mathcal{P}_{\mathrm{i}} + \delta\mathcal{P}_{\mathrm{c}} + \delta\mathcal{P}_{\mathrm{d}} = 0, \tag{2.51}$$

$$\delta\mathcal{P}_{\mathrm{a}} = \int_{\mathcal{V}_t} -\rho\,(\dot{\boldsymbol{v}} \cdot \boldsymbol{f} + \tilde{\kappa})\;\mathrm{d}v, \tag{2.52}$$

$$\delta\mathcal{P}_{\mathrm{d}} = \int_{\mathcal{V}_t} \rho\,(\boldsymbol{b} \cdot \boldsymbol{f} + \beta)\;\mathrm{d}v, \tag{2.53}$$

$$\delta\mathcal{P}_{\mathrm{i}} = \int_{\mathcal{V}_t} -\left(\boldsymbol{\sigma} \cdot \mathrm{grad}\,(\boldsymbol{f}) + l\right)\;\mathrm{d}v, \tag{2.54}$$

$$\delta\mathcal{P}_{\mathrm{c}} = \int_{\partial\mathcal{V}_t} \boldsymbol{t} \cdot \boldsymbol{f} + s\,\mathrm{d}a, \tag{2.55}$$

cf. Forest et al. (2011, pp. 71, 72). Here, $\delta\mathcal{P}_{\mathrm{a}}$ denotes the virtual power of inertial forces, $\delta\mathcal{P}_{\mathrm{i}}$ the virtual power of internal forces, $\delta\mathcal{P}_{\mathrm{c}}$ the virtual power of contact forces, and $\delta\mathcal{P}_{\mathrm{d}}$ the virtual power of far field body forces. The test function is referred to as $\boldsymbol{f}$. If the test function is chosen as virtual displacement, i.e. $\boldsymbol{f} = \delta\boldsymbol{u}$, the PoVW is obtained. Consequently, choosing the test function as virtual velocity, i.e. $\boldsymbol{f} = \delta\boldsymbol{v}$, yields the PoVP. In literature, both the PoVW as well as the PoVP are used to define extended continuum models. The additional contributions κ, β, s and l differ for each considered approach. Regarding a classical Cauchy continuum, these extensions vanish. It appears obvious to relate the additional contributions to DOFs of kinematic nature. Thus, higher-order velocity gradients can be taken into account. For some applications, it is also convenient to supplement a material point by additional, rotational DOFs. Moreover, generic extensions such as a plastic slip, a damage parameter, or an order parameter concerning phase transition are also possible additional DOFs. Subsequently, several extended continua based on their specific extensions are revisited.

Fundamental assumptions

- The primal mechanical principle is that of work, cf. Eugster and dell'Isola (2017, comment on p. 492).

- The notion of stress is presumed from the outset, cf. Eugster and dell'Isola (2017, comment on p. 492).
- The power caused by surface forces is not considered to be fundamental. It is considered as logical consequence arising from the definition of stresses, cf. Eugster and dell'Isola (2017, comment on p. 492).
- Cauchy's lemma is not considered as fundamental concept to derive a continuum mechanical field theory, cf. Eugster and dell'Isola (2017, comment on p. 493).
- After application of partial integration, the localization of volume integrals yields field equations that are commonly referred to as balance equations.
- The material behavior is not specified from the outset.
- To consider a specific material behavior, the dissipation inequality is commonly exploited, cf. Forest (2009). To this end, the definition of both an energy and entropy balance is necessary, in addition, cf. Forest (2005).

2.3.2 Extended principle of virtual power

Micromorphic medium according to Germain (1973) The previously presented micromorphic medium according to Eringen (1999) is treated by means of the principle of virtual power by Germain (1973, pp. 559-566). For brevity, the quasi-static case is presented, here. However, details on a dynamic version can also be found in Germain (1973). The supplementary contributions to the PoVP are given as follows

$$\begin{aligned} \tilde{\kappa} &= 0, & \beta &= \boldsymbol{B}_{\mathrm{E}} \cdot \delta\boldsymbol{\nu}_{\mathrm{E}}, \\ l &= (\boldsymbol{s} - \boldsymbol{\sigma}) \cdot \delta\boldsymbol{\nu}_{\mathrm{E}}^{\mathsf{T}} + \boldsymbol{m}_{\mathrm{E}} \cdot \left(\operatorname{grad}\left(\delta\boldsymbol{\nu}_{\mathrm{E}}^{\mathsf{T}}\right)\right)^{\mathsf{T}_{\mathrm{H}}}, & s &= (\boldsymbol{m}_{\mathrm{E}}[\delta\boldsymbol{\nu}_{\mathrm{E}}]) \cdot \boldsymbol{n}, \end{aligned} \tag{2.56}$$

cf. Germain (1973, pp. 560, 561). The tensors involved correspond to those of Eringen (1999). In contrast to Eq. (2.29), the symmetric second-order tensor $\boldsymbol{s}$ is used in Eq. (2.56). The tensor $\boldsymbol{s}$ is introduced by Eringen in the context of the derivation of the balance of angular momentum, cf. Eringen (1999, p. 44). Thereby, it is noted that s is arbitrary but symmetrical. The micropolar continuum as a special case of the micromorphic continuum is considered by Germain (1973, pp. 570, 571).

Cosserat continuum according to Forest (2005) This continuum is closely related to the micropolar continuum discussed by Kafadar and Eringen (1971); Eringen (1999). According to Eringen (1999, p. 12), the continuum by Cosserat and Cosserat (1909) does not exhibit a micro-inertia. Contrarily, the Cosserat continuum discussed by Forest (2005) provides a micro-inertia. However, a discussion concerning the conservation of micro-inertia is still omitted. The micro-inertia is considered constant. The additional contributions to the PoVP are given by

$$\begin{aligned} \tilde{\kappa} &= \boldsymbol{j} \cdot (\delta\dot{\tilde{\boldsymbol{\nu}}} \otimes \delta\tilde{\boldsymbol{\nu}}), \quad \boldsymbol{j} = \tilde{A}\,\boldsymbol{I}, & \beta &= \tilde{\boldsymbol{B}} \cdot \delta\tilde{\boldsymbol{\nu}}, \\ l &= \boldsymbol{\sigma} \cdot (\boldsymbol{\epsilon}[\delta\tilde{\boldsymbol{\nu}}]) + \tilde{\boldsymbol{m}} \cdot \operatorname{grad}(\delta\tilde{\boldsymbol{\nu}}), & s &= (\tilde{\boldsymbol{m}}\boldsymbol{n}) \cdot \delta\tilde{\boldsymbol{\nu}}, \end{aligned} \tag{2.57}$$

cf. Forest (2005, pp. 4, 5). In this context, the micro-inertia tensor is chosen isotropic, with the micro-inertia constant $\tilde{A}$.

Second gradient continuum according to Germain (1973) The extension by means of the second gradient of the displacement field can be also considered as a particular case of the micromorphic continuum. The connection is drawn in Germain (1973, p. 571).

Indeterminate couple-stress theory according to Germain (1973) This continuum is a special case of the micromorphic continuum. It is obtained by the restriction

$$\boldsymbol{\nu}_{\mathrm{E}} = \operatorname{skw}(\operatorname{grad}(\boldsymbol{v})), \tag{2.58}$$

cf. Germain (1973, Eq. (76)). This theory is also discussed by Fleck and Hutchinson (1997, pp. 337-339).

Formal extension of the principle of virtual power Regarding a quasi-static behavior, a generic extension to the principle of virtual power is given by Forest (2009). This is extended by dynamical contributions in Forest et al. (2011). The corresponding additional terms are given by

$$\begin{aligned} \tilde{\kappa} &= \tilde{A}\ddot{\Phi}\,\delta\dot{\Phi}, & \beta &= a^{\beta}\,\delta\dot{\Phi} + \boldsymbol{b}^{\beta}\cdot\operatorname{grad}\left(\delta\dot{\Phi}\right), \\ l &= a^{l}\,\delta\dot{\Phi} + \boldsymbol{b}^{l}\cdot\operatorname{grad}\left(\delta\dot{\Phi}\right), & s &= a^{s}\,\delta\dot{\Phi}, \end{aligned} \tag{2.59}$$

cf. Forest et al. (2011, pp. 71, 72). Here, the additionally introduced fields are scalar- and vector-valued. However, the tensorial order of the additional fields and thereby the additional DOF is arbitrary, cf. Forest (2009). Thus, the extensions according to Eq. (2.59) describe not a specific continuum but a class of continua. This method is referred to as micromorphic approach. Choosing $\Phi = \chi$, the micromorphic continuum according to Eringen (1999) is obtained, cf. Forest (2009, Eq. (36)). Since the contributions consist of arbitrary fields, the micromorphic approach comprises a variety of generalized continua.

Additional scalar DOF Regarding the micromorphic approach, a special case is given, if the additional DOF is considered to be scalar-valued. By this choice, phenomena such as microstrain gradient plasticity, micro-damage, micro-diffusion, and others can be described, cf. Forest (2009, Tab. 1). If the additional DOF is considered to be the accumulated plastic slip, an accumulated slip gradient crystal plasticity theory can be obtained, cf. Wulfinghoff and Böhlke (2012); Wulfinghoff et al. (2013); Bayerschen et al. (2015; 2016b). Since the field of application is that vast, the specific additional contributions are omitted, here.

Limitations of the methods based on an extended PoVP

- The PoVP does not require the use of Reynolds' transport theorem. This is based on the use of $\tilde{\kappa}$ and not κ in the context of the PoVP. This has the following consequences.

  - The material time derivative of, e.g., the micro-inertia is not discussed. Consequently, no conservation of micro-inertia can be obtained or discussed.
  - Regarding a body that contains a singular surface, the curvature of the singular surface is not intrinsically considered.
- The method provides as many field equations as DOFs are considered, cf. Maugin (1980, p. 64). These are commonly denoted as balance equations. The question, if these field equations are really independent balances or constraints that have to be considered, remains unanswered. The additional field equations are not necessarily coupled to the variables describing the overall deformation of the body.
- In order to model the constitutive material behavior, an extended balance of internal energy along with the dissipation inequality is considered. Therefore, actual but not virtual contributions are required, cf. Hütter (2016, p. 1936). The relation between virtual and actual contributions is commonly not discussed.
- The conservation of mass cannot be proven and is, thus, assumed.
- The existence of the considered stresses is assumed. Regarding not only one but several additional DOFs, the unambiguity of the corresponding stresses cannot be shown.

2.3.3 Extended principle of virtual work

Higher-order velocity gradients The consideration of higher-order gradients of the virtual displacement with respect to the volume-specific work is briefly presented by Hellinger (1913, p. 622). In this context, the second gradient of the virtual displacement is considered in addition. This is motivated by the equations of motion for rods and shells, cf. Hellinger (1913, p. 623). However, only the static version of Eq. (2.55) is considered with respect to an extension, i.e. $\delta\mathcal{P}_{\mathrm{a}} = 0$ holds true. The additional contributions to the PoVW are given by Hellinger (1913, p. 622).

A further discussion of the subsequent exploitation of Eq. (2.55) based on the additional contributions is omitted in the original work. Gradients of the virtual displacement that exceed the second order are considered by Bertram and Forest (2007, p. 9). The original work by Hellinger (1913) is formulated in German. A commented english translation of Hellinger (1913) is given by the articles Eugster and dell'Isola (2017; 2018a;b).

Oriented media Following the idea of Cosserat and Cosserat (1909), an extended, statical principle of virtual work is presented by Hellinger (1913) accounting for the orientation of a material point. Thus, it is assumed that an infinitesimal rotation of the continuum expends virtual work. In this context, the additionally considered extensions are given by

$$\tilde{\kappa} = 0, \quad \beta = \tilde{\boldsymbol{B}} \cdot \delta \boldsymbol{d}, \quad l = \tilde{\boldsymbol{m}} \cdot \mathrm{grad}\,(\delta \boldsymbol{d})\,, \quad s = (\tilde{\boldsymbol{m}}\boldsymbol{n}) \cdot \delta \boldsymbol{d}. \tag{2.60}$$

Here, $\delta \boldsymbol{d}$ denotes an infinitesimal axial vector, cf. Hellinger (1913, p. 623). It can be considered as virtual director field. This is comparable to the additional terms in Eq. (2.57). There is no constraint between the virtual displacement field $\delta \boldsymbol{u}$ and the virtual director field $\delta \boldsymbol{d}$. Thus, both virtual fields can be varied independently of each other. Further discussion of the simplifications concerning two and one dimensional continua can be found in Hellinger (1913). Moreover, a discussion on continua with internal constraints is given there as well.

Higher-order tensorial DOFs as generic extension In Truesdell and Toupin (1960, Section 232), the additional contributions to the PoVW are of generic nature. Thus, they are not related to DOFs that describe the continuum's kinematic in more complexity. The additionally introduced terms consists of work conjugate tensors of second and higher order.

Limitations of the methods based on an extended PoVW The same restrictions apply to the principle of virtual work as to the principle of virtual power.

2.4 Variational Approaches

2.4.1 Fundamentals

Variational principles for extended continua As already mentioned in Section 2.1.1, the principle of virtual power constitutes the stationary condition of the Hamiltonian. Presuming a hyperelastic material behavior for each extended continuum model presented in Section 2.3, a corresponding Lagrange density $\mathcal{L}$ can be formulated that fulfills the principle of least action according to Eq. (2.22). The Lagrangian of an extended continuum depends additionally on the rate and the gradients of the supplementary DOF. Regarding dissipative processes, a dissipation potential has to be taken into account. For brevity, only elastic theories are considered subsequently. It exceeds the work at hand to provide an extensive overview of variational principles. Thus, focus is put on selected works concerning extended continua. A detailed presentation of the Lagrangian of the corresponding continuum is omitted, here.

Fundamental assumptions

- The Lagrangian accounts for a specific material behavior.
- The equations of motion associated with the corresponding DOF are obtained as stationary condition of the Hamiltonian.
- In its original formulation, the principle of least action does not account for dissipative processes.

2.4.2 Extended continua specified by an extended Lagrangian

Cosserat continuum This continuum is originally proposed by Cosserat and Cosserat (1909). The corresponding Lagrange density is formulated on p. 4 of the extensive treatment. Similar to this, a body with oriented media is considered by Hellinger (1913).

Continua with couple-stress This special case of the Cosserat continuum, cf. Germain (1973), is treated by Mindlin and Tiersten (1962); Toupin (1962; 1964) in a variational context. An error in the equation of the couple-stress, given in Truesdell and Toupin (1960), is discussed. The essay by Toupin (1964) constitutes a review on different continua with couple-stresses.

Microdeformation Similar to Eringen's extended continuum, Mindlin (1964) introduces a micro-displacement related to a micro-volume included in a material point. Materials with a complex microstructure are considered by Capriz and Mariano (2003). The material time derivative of the specific internal energy is used as Lagrangian by Rahouadj et al. (2003).

Displacement gradients of higher order The continuum according to Agiasofitou and Lazar (2009) considers a linear elastic continuum of grade three. To this end, the Lagrangian depends on the second gradient of the displacement field.

Limitations of variational approaches

- The form of the Lagrangian depends on the constitutive assumptions. Therefore, the validity of the preserved field equations is very restricted.
- Accounting for dissipative processes can be quite involved.

2.5 Postulates

In order to derive balance equations of generalized continua, several procedures are available as discussed previously. Mariano (2016) lists ten approaches that are admissible from his point of view. This canon of methods also considers the postulation of balance equations in a local or an integral form. In this context, he lists the disadvantages of this procedure. Moreover, the postulation of an additional microstructural balance equation in integral form is critically discussed by Mariano (2016, p. 14).

In addition, the postulation of arbitrary additional balance equations is also questioned in Yavari and Marsden (2009, p. 7). Nevertheless, several authors postulate appropriate balance laws suitable for their field of application. Some of the publications that directly postulate balance equations are given along with the number of the corresponding equation: Leslie (1968, Eq. (3.2)), Naghdi and Srinivasa (1993, Eq. (4.10), Eq. (4.11)), Gurtin (1993, Eq. (10.2)), Gurtin (1994, Eq. (2.2)), Gurtin (1995, Eq. (3.23)), Fried (1996, Eq. (3.2)), Rahaman et al. (2016, (2.16)), Rahaman et al. (2017, Eq. (2.12)).

2.6 Interim Conclusions

The following results and concluding remarks are given with respect to the current chapter.

- Methods are presented that are used to obtain balance equations associated with additional degrees of freedom. In this context, various extended continua are outlined.
- Invariance considerations of an extended energy balance as well as an extended principle of virtual work do not consider material behavior. They are therefore not restricted in their application.
- Both the covariance of an extended energy balance as well as variational principles take into account constitutive equations describing the material behavior a priori. Consequently, obtained balance laws are always associated with a specific material behavior.
- Some publications directly postulate additional balance equations which are in line with balance equations obtained by the previously mentioned methods. However, this approach induces a certain degree of arbitrariness.
- In order to obtain thermodynamically consistent constitutive equations, the second law of thermodynamics has to be exploited. Regarding an

extended energy balance, supplementary contribution enter the dissipation inequality by means of the Legendre transformation, cf. Beegle et al. (1974). In the context of an extended principle of virtual power, an extended balance of internal energy has to be postulated in addition, cf. Forest (2009). Thus, actual supplementary contributions have to be postulated alongside the virtual contributions, cf. Hütter (2016). Consequently, the principle of virtual power is not self-consistent in deriving a closed theory.

- This motivates the question of the relationship between the modified equations obtained by an extended energy balance and the additional balance equations obtained by an extended principle of virtual power.

The relationship between the previously outlined theories concerning the invariance of an extended energy balance and an extended PoVP is illustrated in Fig. 2.2.

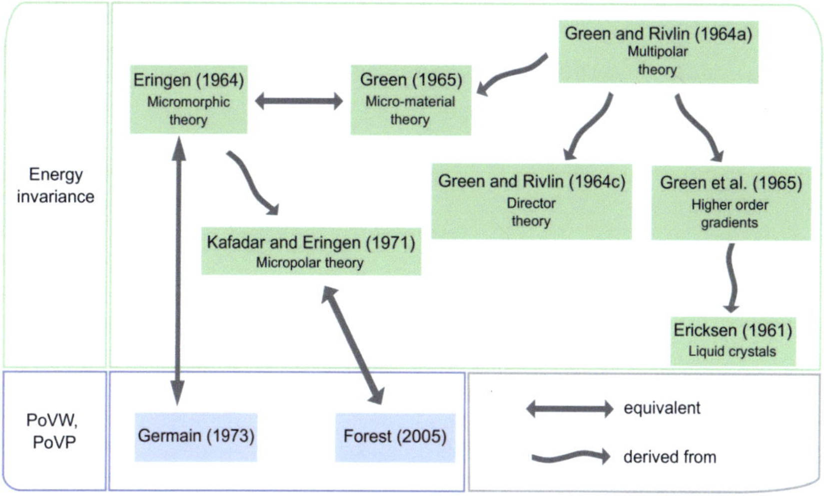

Figure 2.2: This graphic illustrates the relation between the extended continua previously outlined. Equivalent theories are indicated by a straight double-headed arrow. The wavy arrow indicates theories that are obtained by means of simplifications or specifications of the additional contributions of the energy balance.

Chapter 3

On Invariance Properties of an Extended Energy Balance[1]

Motivation Gradient plasticity theories are of utmost importance for accounting for size effects in metals, especially on the grain scale. Today, there are several methods used to derive the governing equations for the additional degrees of freedom in gradient plasticity theories. Here, the equivalence between an extended principle of virtual power and an extended energy balance is shown. The energy balance of a Boltzmann continuum is supplemented by contributions based on a scalar-valued degree of freedom. It is considered to be invariant with respect to a change of observer. This yields unambiguously the existence of a corresponding gradient stress vector, which is presumed from the outset in the context of an extended principle of virtual power. A thermodynamically consistent nonlocal evolution equation for the additional, scalar-valued degree of freedom is obtained by evaluation of the dissipation inequality in terms of the Clausius-Duhem inequality. Partitioning the nonlocal flow rule yields a partial differential equation, often referred to as micro-force balance. Moreover, a discussion concerning an additional vectorial DOF and the derivation of the corresponding conservation of micro-inertia is also provided.

[1] The content of this chapter is taken directly from the article Prahs and Böhlke (2019b). Minor linguistic changes and abbreviations have been made.

3.1 Additional Scalar-Valued DOF

3.1.1 Extended energy balance

Deformation and microstructure function The energy balance according to Eq. (2.28) is given with respect to a current configuration referred to as $\mathcal{C}$. The spatial velocity field $\boldsymbol{v}$ is given by Eq. (2.1). Subsequently, the additional DOF is referred to as γ. The spatial velocity $\tilde{v}$ of γ is calculated according to Eq. (2.27).

Energy balance The considered extensions to the energy balance, based on the additional DOF, are given by Eq. (2.43). Thus, the extended energy balance according to Eq. (2.28) can be written in the form

$$\begin{aligned}\epsilon = \int_{\mathcal{V}_t} & (\dot{\rho} + \rho \mathrm{div}\,(\boldsymbol{v})) \left(e + \frac{1}{2}\boldsymbol{v}\cdot\boldsymbol{v} + \frac{1}{2}\tilde{A}\tilde{v}^2 \right) + \rho\dot{e} + \rho\,(\boldsymbol{a} - \boldsymbol{b})\cdot\boldsymbol{v} \\ & + \rho\left(\tilde{A}\tilde{a} - \tilde{b}\right)\tilde{v} + \frac{1}{2}\rho\dot{\tilde{A}}\tilde{v}^2 - \rho r \,\mathrm{d}v - \int_{\partial\mathcal{V}_t} \boldsymbol{t}\cdot\boldsymbol{v} + \tilde{t}\tilde{v} + h\,\mathrm{d}a = 0. \qquad (3.1)\end{aligned}$$

As discussed in Sec. 2.1.1, it is assumed that the energy balance is invariant with respect to a change of observer, described by Eq. (2.13). The additionally considered DOF is scalar-valued and, thus, not affected by a change of observer. The energy balance described by the observer of the vector space $\mathcal{W}$ is denoted by ϵ. Consequently, the observer of the vector space $\mathcal{W}'$ refers to the energy balance as ϵ'. To obtain the balance laws, the difference

$$\Delta\epsilon_{t_0} = (\epsilon' - \epsilon)|_{t=t_0} \qquad (3.2)$$

is introduced and evaluated. The considered calculations are closely related to the discussion given in Marsden and Hughes (1994, pp. 145-149). Instead of the Euclidean transformation according to Eq. (2.13), it is also possible to consider a pure translational transformation first and a pure rotational one subsequently.

3.1.2 Translational transformation

Applied transformations Regarding a pure translational transformation implies that both observers exhibit the same orientation, i.e., $\boldsymbol{Q} = \boldsymbol{e}'_i \otimes \boldsymbol{e}_i$. Consequently, the Euclidean transformation according to Eq. (2.13) is simplified. The relations between $\boldsymbol{x}$ and $\boldsymbol{x}'$,$\boldsymbol{v}$ and $\boldsymbol{v}'$, as well as $\boldsymbol{a}$ and $\boldsymbol{a}'$ are given by

$$\boldsymbol{x}' = \boldsymbol{x} + \boldsymbol{c}', \quad \boldsymbol{v}' = \boldsymbol{v} + \boldsymbol{w}, \quad \boldsymbol{a}' = \boldsymbol{a} + \dot{\boldsymbol{w}}, \tag{3.3}$$

with $\boldsymbol{w} = \dot{\boldsymbol{c}}'$. All scalar quantities, including the contributions associated with the additional DOF, are invariant under the considered transformations. Moreover, the surface traction $\boldsymbol{t}$ remains unchanged under the translational transformation. Regarding the body force $\boldsymbol{b}'$, an additional contribution associated with the acceleration $\ddot{\boldsymbol{c}}'$ of the relative translation has to be taken into account. This contribution is referred to as fictitious body force, cf. Marsden and Hughes (1994, p. 147). Consequently, the transformation

$$\boldsymbol{a}' - \boldsymbol{b}' = \boldsymbol{a} - \boldsymbol{b} \tag{3.4}$$

holds true. Alternatively, the requirement of $\ddot{\boldsymbol{c}}' = \boldsymbol{0}$ also ensures the validity of Eq. (3.4), cf. Yavari and Marsden (2009, p. 9) and Marsden and Hughes (1994, p. 146). As a consequence of the transformations according to Eq. (3.3), the kinetic contribution to the energy $\boldsymbol{v} \cdot \boldsymbol{v}/2$ is not invariant under the translational transformation. It contains additional terms that are linear and quadratic in $\dot{\boldsymbol{c}}'$. The arbitrariness of $\boldsymbol{c}'$ and, thus, $\dot{\boldsymbol{c}}'$ is used to derive the balance of mass, subsequently. This is an essential aspect for further invariance considerations of the extended energy balance.

Existence of the Cauchy stress tensor Evaluation of Eq. (4.16) yields

$$\Delta\epsilon_{t_0} = \int_{\mathcal{V}_t} (\dot{\rho} + \rho \operatorname{div}(\boldsymbol{v})) \left(\boldsymbol{v} \cdot \boldsymbol{w} + \frac{1}{2} \boldsymbol{w} \cdot \boldsymbol{w} \right) + \rho \, (\boldsymbol{a} - \boldsymbol{b}) \cdot \boldsymbol{w} \, \mathrm{d}v - \int_{\partial \mathcal{V}_t} \boldsymbol{t} \cdot \boldsymbol{w} \, \mathrm{d}a = 0. \tag{3.5}$$

Application of Eq. (3.5) to an infinitesimal tetrahedron yields the existence of the Cauchy stress tensor $\boldsymbol{\sigma}$

$$\boldsymbol{\sigma} \boldsymbol{n} = \boldsymbol{t}, \tag{3.6}$$

cf. Bertram (2005, p. 138).

Conservation of mass Accounting for Eq. (3.6), Eq. (3.5) can be formulated as

$$\Delta\epsilon_{t_0} = \int_{\mathcal{V}_t} (\dot{\rho} + \rho \operatorname{div}(\boldsymbol{v})) \left(\boldsymbol{v} \cdot \boldsymbol{w} + \frac{1}{2} \boldsymbol{w} \cdot \boldsymbol{w} \right) + (\rho \, (\boldsymbol{a} - \boldsymbol{b}) - \operatorname{div}(\boldsymbol{\sigma})) \cdot \boldsymbol{w} - \boldsymbol{\sigma} \cdot \operatorname{grad}(\boldsymbol{w}) \, \mathrm{d}v = 0. \tag{3.7}$$

The vector field $\boldsymbol{w}$ is arbitrary and given by $\boldsymbol{w} = \dot{\boldsymbol{c}}'$. Thus, $\boldsymbol{w} = \lambda \boldsymbol{u}$ is considered, with the constant unit vector $\boldsymbol{u}$ and $\lambda \neq \lambda(\boldsymbol{x})$, cf. Marsden and Hughes (1994, p. 148). This leads to

$$\Delta\epsilon_{t_0} = \int_{\mathcal{V}_t} (\dot{\rho} + \rho \operatorname{div}(\boldsymbol{v})) \left(\lambda \boldsymbol{v} \cdot \boldsymbol{u} + \frac{1}{2} \lambda^2 \boldsymbol{u} \cdot \boldsymbol{u} \right) + \lambda \, (\rho \, (\boldsymbol{a} - \boldsymbol{b}) - \operatorname{div}(\boldsymbol{\sigma})) \cdot \boldsymbol{u} \, \mathrm{d}v = 0. \tag{3.8}$$

Differentiating Eq. (3.8) twice with respect to lambda yields

$$\frac{d^2 \Delta\epsilon_{t_0}}{d\lambda^2} = \int_{\mathcal{V}_t} (\dot{\rho} + \rho \operatorname{div}(\boldsymbol{v})) \, (\boldsymbol{u} \cdot \boldsymbol{u}) \, \mathrm{d}v = 0. \tag{3.9}$$

Since $\boldsymbol{u}$ is a constant unit vector, i.e., $\boldsymbol{u} \cdot \boldsymbol{u} = 1$, conservation of mass is obtained in its local form, reading

$$\dot{\rho} + \rho \operatorname{div}(\boldsymbol{v}) = 0. \tag{3.10}$$

Balance of linear momentum Accounting for conservation of mass, cf. Eq. (3.10), Eq. (3.8) can be formulated as

$$\Delta \epsilon_{t_0} = \int_{\mathcal{V}_t} \lambda \left(\rho \left(\boldsymbol{a} - \boldsymbol{b}\right) - \operatorname{div}(\boldsymbol{\sigma})\right) \cdot \boldsymbol{u} \, \mathrm{d}v = 0. \tag{3.11}$$

Since λ and $\boldsymbol{u}$ are arbitrary, localization of Eq. (3.11) yields the local form of the balance of linear momentum as

$$\rho \left(\boldsymbol{a} - \boldsymbol{b}\right) - \operatorname{div}(\boldsymbol{\sigma}) = 0. \tag{3.12}$$

3.1.3 Rotational transformation

Applied transformations Considering a pure rotational transformation implies that both observers share the same origin, i.e., $\boldsymbol{c}' = \boldsymbol{0}$. Consequently, the relation between $\boldsymbol{x}$ and $\boldsymbol{x}'$, respectively, between $\boldsymbol{v}$ and $\boldsymbol{v}'$ is given by

$$\boldsymbol{x}' = \boldsymbol{Q}\boldsymbol{x}, \quad \boldsymbol{v}' = \boldsymbol{Q}\boldsymbol{v} + \boldsymbol{w}, \tag{3.13}$$

with $\boldsymbol{w} = \dot{\boldsymbol{Q}}\boldsymbol{x}$. The traction force is assumed to transform objectively, reading

$$\boldsymbol{t}' = \boldsymbol{Q}\boldsymbol{t}. \tag{3.14}$$

Contrarily, the acceleration $\boldsymbol{a}$ does not transform objectively, which is a consequence of the transformation law for $\boldsymbol{v}'$, cf. Eq. (3.13). Additional contributions associated with the centripetal and Coriolis forces are added.

Taking into account these fictitious forces, the difference between body force and acceleration is assumed to transform objectively

$$\left(\boldsymbol{a}' - \boldsymbol{b}'\right) = \boldsymbol{Q}\left(\boldsymbol{a} - \boldsymbol{b}\right), \tag{3.15}$$

cf. Marsden and Hughes (1994). It is assumed that $\boldsymbol{Q}(t_0) = \boldsymbol{I}$ holds true. According to the assumptions of the previous section, all scalar quantities are considered invariant with respect to the applied transformations. This implies that p, $\tilde{t}$, $\tilde{b}$ and $\tilde{v}$ are invariant with respect to the rotational transformations.

Conservation of micro-inertia Evaluation of Eq. (4.16) leads to a form that does not contain any microstructural quantities associated with the additional DOFs, reading

$$\Delta\epsilon_{t_0} = \int_{\mathcal{V}_t} -\boldsymbol{\sigma} \cdot \operatorname{grad}\left(\boldsymbol{w}\right) \,\mathrm{d}v = 0. \tag{3.16}$$

Thus, the existence of a micro-inertia conservation cannot be proved based on invariance considerations regarding a change of observer. The same holds true if invariance with respect to a superimposed rigid-body motion is considered. This is different from the consideration of an additional vectorial DOF, cf. Eq. (3.58).

Balance of angular momentum Substituting $\boldsymbol{w}$ according to Eq. (3.13) into Eq. (3.16) yields

$$\Delta\epsilon_{t_0} = \int_{\mathcal{V}_t} -\boldsymbol{\sigma} \cdot \dot{\boldsymbol{Q}} \,\mathrm{d}v = 0. \tag{3.17}$$

Since $\dot{\boldsymbol{Q}}(t_0) \in \mathit{Skw}$, the localization of Eq. (3.17) yields the standard balance of angular momentum, as given in Eq. (2.15), reading

$$\boldsymbol{\sigma} = \boldsymbol{\sigma}^{\mathsf{T}}. \tag{3.18}$$

Existence of the gradient stress vector Accounting for the results in Eqs. (3.6), (3.10) and (3.12), Eq. (3.1) reads

$$\int_{\mathcal{V}_t} \rho\dot{e} + \rho\left(\tilde{A}\tilde{a} - \tilde{b}\right)\tilde{v} + \frac{1}{2}\rho\dot{\tilde{A}}\tilde{v}^2 - \rho r - \boldsymbol{\sigma}\cdot \operatorname{grad}\left(\boldsymbol{v}\right)\,\mathrm{d}v$$
$$- \int_{\partial\mathcal{V}_t} \tilde{t}\tilde{v} + h\,\mathrm{da} = 0. \tag{3.19}$$

Application of Eq. (3.19) to an infinitesimal tetrahedron yields

$$\int_{\partial\mathcal{V}_t} \tilde{t}\tilde{v} + h\,\mathrm{da} = 0. \tag{3.20}$$

The integrand of the surface integral consists of the contribution due to the micro surface traction $\tilde{t}$ and the heat flux h. The existence of a flux term $\boldsymbol{k}(\boldsymbol{x},t)$ can be proven for which

$$\boldsymbol{k}(\boldsymbol{x},t)\cdot\boldsymbol{n} = \tilde{t}(\boldsymbol{x},t,\boldsymbol{n})\,\tilde{v}(\boldsymbol{x},t) + h(\boldsymbol{x},t,\boldsymbol{n}) \tag{3.21}$$

holds, cf. Marsden and Hughes (1994, p. 127). Both, $\tilde{t}$ and h do not depend on $\tilde{v}$. Thus, the only possible choice for $\boldsymbol{k}(\boldsymbol{x},t)$ that provides the integrand of Eq. (3.20) is given by

$$\boldsymbol{k}(\boldsymbol{x},t) = \boldsymbol{\xi}\tilde{v} + \boldsymbol{h}, \quad \text{with } \tilde{t} = \boldsymbol{\xi}\cdot\boldsymbol{n} \quad \text{and } h = \boldsymbol{h}\cdot\boldsymbol{n}. \tag{3.22}$$

Consequently, the existence of the gradient stress vector $\boldsymbol{\xi}$ is shown unambiguously. This is different from the treatment of extended continua by an extended principle of virtual power. In this context, the existence of both the Cauchy stress tensor and the gradient stress vector is presumed from the outset, cf. the remark on this topic in the review paper of Mariano (2016, p. 14).

Existence of heat flux vector Regarding Eq. (3.22), the heat flux vector $\boldsymbol{q}$ is introduced such that

$$\boldsymbol{q} = -\boldsymbol{h}, \quad \text{with } \boldsymbol{q} \cdot \boldsymbol{n} = -h, \tag{3.23}$$

cf. Marsden and Hughes (1994, p. 148).

Balance of internal energy Considering the results in Eqs. (3.6), (3.10), (3.12), (3.18) and (3.22), localization of Eq. (3.1) yields the local form of the balance of internal energy, reading

$$\begin{aligned} &\rho\dot{e} + \rho\left(\tilde{A}\tilde{a} - \tilde{b}\right)\tilde{v} + \frac{1}{2}\rho\dot{\tilde{A}}\tilde{v}^2 - \rho r - \boldsymbol{\sigma}\cdot\boldsymbol{D} \\ &- \boldsymbol{\xi}\cdot\operatorname{grad}\left(\tilde{v}\right) - \operatorname{div}\left(\boldsymbol{\xi}\right)\tilde{v} + \operatorname{div}\left(\boldsymbol{q}\right) = 0. \end{aligned} \tag{3.24}$$

Simplifying assumptions Additionally considered DOFs are commonly used to describe the evolution of the underlying microstructure. Nonlocal damage, cf. Germain et al. (2007), nonlocal diffusion, cf. Ubachs et al. (2004) and nonlocal plasticity, cf. Wulfinghoff et al. (2013), are prominent examples for the application of extended continua. In this context, effects due to micro-inertia and micro-body forces are usually neglected, i.e., $\tilde{A} = 0$ and $\tilde{b} = 0$. Only micro-traction forces are considered. Thus, the balance of internal energy Eq. (3.24) reads

$$\rho\dot{e} - \rho r - \boldsymbol{\sigma}\cdot\boldsymbol{D} - \boldsymbol{\xi}\cdot\operatorname{grad}\left(\tilde{v}\right) - \operatorname{div}\left(\boldsymbol{\xi}\right)\tilde{v} + \operatorname{div}\left(\boldsymbol{q}\right) = 0. \tag{3.25}$$

3.1.4 Nonlocal evolution equation for an additional DOF

Exploitation of the Clausius-Duhem inequality To discuss the evolution equation for the additional DOF, the simplified balance of internal energy according to Eq. (3.25) is considered, subsequently. Moreover, a small strain framework is considered for brevity, i.e., $\boldsymbol{D} = \dot{\boldsymbol{\varepsilon}}$ holds. Consequently, the Clausius-Duhem inequality according to Eq. (2.19) is given by

$$\rho\delta = \boldsymbol{\sigma} \cdot \dot{\boldsymbol{\varepsilon}} + \boldsymbol{\xi} \cdot \nabla\dot{\gamma} + \mathrm{div}\,(\boldsymbol{\xi})\,\dot{\gamma} - \rho\dot{\psi} - \rho\dot{\theta}\eta - \frac{1}{\theta}\boldsymbol{q} \cdot \boldsymbol{g} \geq 0, \tag{3.26}$$

where $\dot{\gamma} = \tilde{v}$ and $\nabla\dot{\gamma} = \mathrm{grad}\,(\dot{\gamma})$ is used. An additive split of the infinitesimal strain $\boldsymbol{\varepsilon}$ into a purely elastic part $\boldsymbol{\varepsilon}^{\mathrm{e}}$ and a part $\boldsymbol{\varepsilon}^{\mathrm{p}}(\gamma)$ related to the additional DOF γ is assumed, i.e., $\boldsymbol{\varepsilon} = \boldsymbol{\varepsilon}^{\mathrm{e}} + \boldsymbol{\varepsilon}^{\mathrm{p}}$ holds true. The specific free energy is assumed to depend on $\boldsymbol{\varepsilon}, \boldsymbol{\varepsilon}^{\mathrm{p}}, \gamma, \nabla\gamma, \theta$, i.e.,

$$\psi = \psi(\boldsymbol{\varepsilon}, \boldsymbol{\varepsilon}^{\mathrm{p}}, \gamma, \nabla\gamma, \theta) \tag{3.27}$$

holds true. It is assumed that the elastic properties are not affected by $\boldsymbol{\varepsilon}^{\mathrm{p}}$ during the deformation process, similar to Bertram and Krawietz (2012, p. 2262). This motivates that ψ only depends on the elastic strain $\boldsymbol{\varepsilon}^{\mathrm{e}} = \boldsymbol{\varepsilon} - \boldsymbol{\varepsilon}^{\mathrm{p}}$. Furthermore, for simplicity, it is assumed that the specific free energy ψ can be additively decomposed into an elastic contribution ψ_{e}, a contribution ψ_{h} that depends on the additional DOF γ, a gradient contribution ψ_{g} that accounts for the effects of the gradient of the additional DOF, and a thermal contribution ψ_θ, i.e.,

$$\psi(\boldsymbol{\varepsilon} - \boldsymbol{\varepsilon}^{\mathrm{p}}, \gamma, \nabla\gamma, \theta) = \psi_{\mathrm{e}}(\boldsymbol{\varepsilon} - \boldsymbol{\varepsilon}^{\mathrm{p}}) + \psi_{\mathrm{h}}(\gamma) + \psi_{\mathrm{g}}(\nabla\gamma) + \psi_\theta(\theta). \tag{3.28}$$

Naturally, this assumed split of the specific free energy represents a special case, cf. Bertram and Krawietz (2012). Regarding rate-dependent material behavior, the Clausius-Duhem inequality, cf. Eq. (3.26), reads

$$\begin{aligned}\rho\delta = {} & \left(\boldsymbol{\sigma} - \rho\frac{\partial\psi_{\mathrm{e}}}{\partial\boldsymbol{\varepsilon}}\right) \cdot \dot{\boldsymbol{\varepsilon}} - \rho\frac{\partial\psi_{\mathrm{e}}}{\partial\boldsymbol{\varepsilon}^{\mathrm{p}}} \cdot \frac{\partial\boldsymbol{\varepsilon}^{\mathrm{p}}}{\partial\gamma}\dot{\gamma} - \rho\left(\eta + \frac{\partial\psi_\theta}{\partial\theta}\right)\dot{\theta} - \boldsymbol{q} \cdot \boldsymbol{g}/\theta \\ & + \left(\mathrm{div}\,(\boldsymbol{\xi}) - \rho\frac{\partial\psi_{\mathrm{h}}}{\partial\gamma}\right)\dot{\gamma} + \left(\boldsymbol{\xi} - \rho\frac{\partial\psi_{\mathrm{g}}}{\partial\nabla\gamma}\right) \cdot \nabla\dot{\gamma} \geq 0.\end{aligned} \tag{3.29}$$

The standard procedure of Coleman and Noll is applied, cf. Coleman and Noll (1963). It is assumed that the gradient stress $\boldsymbol{\xi}$ is purely energetic.

This yields the potential relations for the Cauchy stress, the entropy and the generalized stress

$$\boldsymbol{\sigma} = \rho \frac{\partial \psi_{\mathrm{e}}}{\partial \boldsymbol{\varepsilon}}, \quad \eta = -\frac{\partial \psi_{\theta}}{\partial \theta}, \quad \boldsymbol{\xi} = \rho \frac{\partial \psi_{\mathrm{g}}}{\partial \nabla \gamma}. \tag{3.30}$$

Thus, the reduced dissipation inequality is given by

$$\left(\mathrm{div}\,(\boldsymbol{\xi}) - \rho \frac{\partial \psi_{\mathrm{h}}}{\partial \gamma} - \rho \frac{\partial \psi_{\mathrm{e}}}{\partial \boldsymbol{\varepsilon}^{\mathrm{p}}} \cdot \frac{\partial \boldsymbol{\varepsilon}^{\mathrm{p}}}{\partial \gamma} \right) \dot{\gamma} - \boldsymbol{q} \cdot \boldsymbol{g} / \theta \geq 0. \tag{3.31}$$

While the first term of Eq. (3.31) refers to the mechanical dissipation, the thermal dissipation is represented by the second expression.

Evolution equation Subsequently, no coupling is assumed between the mechanical and the thermal dissipation. Thus, Fourier's law, cf. Bertram (2015), ensures the positivity of the second term in Eq. (3.31). Linear irreversible thermodynamics yields an admissible choice for $\dot{p}$, consistent with the reduced dissipation inequality, reading

$$\dot{\gamma} = \dot{\gamma}_0 \left(\mathrm{div}\,(\boldsymbol{\xi}) - \rho \frac{\partial \psi_{\mathrm{h}}}{\partial \gamma} - \rho \frac{\partial \psi_{\mathrm{e}}}{\partial \boldsymbol{\varepsilon}^{\mathrm{p}}} \cdot \frac{\partial \boldsymbol{\varepsilon}^{\mathrm{p}}}{\partial \gamma} \right), \quad \text{with } \dot{\gamma}_0 \geq 0. \tag{3.32}$$

Here, $\dot{\gamma}_0$ denotes a referential rate. Equation (3.32) constitutes a nonlocal evolution equation for the additionally considered DOF. Partitioning of Eq. (3.32) leads to a partial differential equation (PDE) and a local evolution equation, given by

$$\pi - \mathrm{div}\,(\boldsymbol{\xi}) = 0, \quad \dot{\gamma} = \dot{\gamma}_0 \left(\pi - \rho \frac{\partial \psi_{\mathrm{h}}}{\partial \gamma} - \rho \frac{\partial \psi_{\mathrm{e}}}{\partial \boldsymbol{\varepsilon}^{\mathrm{p}}} \cdot \frac{\partial \boldsymbol{\varepsilon}^{\mathrm{p}}}{\partial \gamma} \right). \tag{3.33}$$

For vanishing rates, i.e., $\dot{\gamma} = 0$, the nonlocal evolution equation according to Eq. (3.32) reduces to a partial differential equation, reading

$$\mathrm{div}\,(\boldsymbol{\xi}) - \rho \frac{\partial \psi_{\mathrm{h}}}{\partial \gamma} - \rho \frac{\partial \psi_{\mathrm{e}}}{\partial \boldsymbol{\varepsilon}^{\mathrm{p}}} \cdot \frac{\partial \boldsymbol{\varepsilon}^{\mathrm{p}}}{\partial \gamma} = 0. \tag{3.34}$$

Equation (3.34) characterizes the distribution of the additional DOF γ in thermodynamical equilibrium. The same result is obtained by requiring an isothermal behavior along with a homogeneous temperature distribution and neglecting any effects associated with an energy supply, cf. Eq. (3.39). This is outlined, subsequently.

Vanishing dissipation in regular points Applying the material time derivative to the Legendre transformation yields $\dot{e} = \dot{\psi} + \dot{\theta}\eta + \theta\dot{\eta}$. Since $\psi_\theta = \psi_\theta(\theta)$ and $\eta = -\partial\psi_\theta/\partial\theta$, cf. Eq. (3.30), the material time derivative of the specific entropy reads $\dot{\eta} = \dot{\theta}\partial\eta/\partial\theta = -\dot{\theta}\partial^2\psi_\theta/\partial\theta^2$. Consequently, the material time derivative of the Legendre transformation can be written as

$$\dot{e} = \dot{\psi} + \dot{\theta}\left(\eta + \theta\partial\eta/\partial\theta\right). \tag{3.35}$$

Thus, $\dot{e} = \dot{\psi}$ holds true for an isothermal material behavior, i.e., $\dot{\theta} = 0$, which is commonly assumed in the context of plasticity, cf. Hochrainer (2016, p. 17). Accounting for a homogeneous temperature distribution leads to $\boldsymbol{g} = \operatorname{grad}(\theta) \equiv \boldsymbol{0}$ and thereby $\boldsymbol{q} = \boldsymbol{0}$ by means of Fourier's law, cf., e.g., Bertram (2015). Furthermore, any effects due to heat supply are neglected, i.e., $r = 0$. Thus, the balance of internal energy Eq. (3.25) reads

$$\rho\dot{\psi} - \boldsymbol{\sigma}\cdot\dot{\boldsymbol{\varepsilon}} - \left(\boldsymbol{\xi}\cdot\nabla\dot{\gamma} + \operatorname{div}(\boldsymbol{\xi})\,\dot{\gamma}\right) = 0. \tag{3.36}$$

Moreover, the dissipation inequality according to Eq. (3.26) is given by

$$\rho\delta = -\rho\dot{\psi} + \boldsymbol{\sigma}\cdot\dot{\boldsymbol{\varepsilon}} + \left(\operatorname{div}(\boldsymbol{\xi})\,\dot{\gamma} + \boldsymbol{\xi}\cdot\nabla\dot{\gamma}\right) \geq 0. \tag{3.37}$$

Comparison of Eq. (3.36) and Eq. (3.37) reveals, that the dissipation vanishes under the met assumptions. Regarding small deformations and accounting for Eq. (3.28) with vanishing thermal contriubtion as well as for Eq. (3.30), the balance of internal energy according to Eq. (3.36) reads

$$\left(\operatorname{div}(\boldsymbol{\xi}) - \rho\frac{\partial\psi_{\mathrm{h}}}{\partial\gamma} + \tau\right)\dot{\gamma} = 0. \tag{3.38}$$

The equation above is fullfilled for arbitrary $\dot{\gamma}$, if the terms in brackets vanish. This leads to an additional field equation as already given by Eq. (3.34), often referred to as micro-force balance

$$\operatorname{div}(\boldsymbol{\xi}) - \rho\frac{\partial\psi_{\mathrm{h}}}{\partial\gamma} + \tau = 0, \tag{3.39}$$

cf. Bayerschen et al. (2016b); Gurtin (2002; 2008); Wulfinghoff et al. (2013). Consequently, Eq. (3.34) does not only describe the distribution of the additional DOF in thermodynamical equilibrium, but also for arbitrary rates of the DOF, if the mentioned restrictions are considered.

3.1.5 Connection to an extended principle of virtual power

Weak forms Subsequently, the connection of the presented framework to an extended principle of virtual work is discussed. To this end, the weak forms of the PDE according to Eq. $(3.33)_1$ and the balance of linear momentum Eq. (3.12) are provided, first. Multiplication of Eq. $(3.33)_1$ with a test function f, integration over $\mathcal{V}_t$ and application of the divergence theorem yield the corresponding weak form

$$-\int_{\mathcal{V}_t} \pi f + \boldsymbol{\xi}\cdot\operatorname{grad}(f)\,\mathrm{d}v + \int_{\partial\mathcal{V}_t} \tilde{t} f\,\mathrm{d}a = 0, \tag{3.40}$$

with $\tilde{t} = \boldsymbol{\xi}\cdot\boldsymbol{n}$. Moreover, the weak form of the balance of linear momentum in Eq. (3.12) is given by,

$$\int_{\mathcal{V}_t} \rho(\boldsymbol{a}-\boldsymbol{b})\cdot\boldsymbol{f} + \boldsymbol{\sigma}\cdot\operatorname{grad}(\boldsymbol{f})\,\mathrm{d}v - \int_{\partial\mathcal{V}_t} \boldsymbol{t}\cdot\boldsymbol{f}\,\mathrm{d}a = 0, \tag{3.41}$$

where $\boldsymbol{f}$ is the vectorial test function.

Extended principle of virtual power Subsequently, the quasi-static case is considered and body forces are neglected, i.e., $\boldsymbol{a} = \mathbf{0}$ and $\boldsymbol{b} = \mathbf{0}$. Replacing the test functions f and $\boldsymbol{f}$ by the virtual rates $\delta\tilde{v}$ and $\delta\boldsymbol{v}$, respectively, the sum of Eq. (4.43) and Eq. (4.40) yields $\delta\mathcal{P}_{\text{int}} = \delta\mathcal{P}_{\text{ext}}$, with

$$\delta\mathcal{P}_{\text{int}} = \int_{\mathcal{V}_t} \pi\delta\tilde{v} + \boldsymbol{\xi}\cdot\text{grad}\,(\delta\tilde{v}) + \boldsymbol{\sigma}\cdot\text{grad}\,(\delta\boldsymbol{v})\,\text{d}v,$$

$$\delta\mathcal{P}_{\text{ext}} = \int_{\partial\mathcal{V}_t} \tilde{t}\delta\tilde{v} + \boldsymbol{t}\cdot\delta\boldsymbol{v}\,\text{d}a \qquad (3.42)$$

denoting the internal and external virtual power, respectively. After exploitation of the Coleman-Noll procedure, the set of equations obtained by this extended principle of virtual power is equivalent to the set of equations obtained by an extended energy balance, cf. Section A.1. With respect to isothermal processes, the equivalence between both approaches is mentioned in Svendsen (2011, p. 14). In this context, reference is made to the nondissipative continua discussed by Toupin (1964). The application of an extended principle of virtual power is widely used to derive additional field equations regarding extended continua, cf. Forest (2009). The extended principle of virtual power using Eq. (3.42) is structurally equivalent to Wulfinghoff et al. (2013, Eqs. (3) and (4)), Bayerschen and Böhlke (2016, Eqs. (3) and (4)), similar to, e.g., Cermelli and Gurtin (2002, Eq. (3.2)), Gurtin et al. (2007, Eq. (3.2)). In this context, Eq. $(3.33)_1$ is referred to as additional balance equation. It is denoted as *micro-force balance*, however, the notion of a balance is misleading according to the previous discussions. It is rather part of a partitioned, time-dependent partial differential equation.

3.2 Additional Vector-Valued DOF

3.2.1 Extended energy balance

Subsequently, the additionally considered DOF is referred to as $\boldsymbol{p}$ and defined on the tangent space of the Euclidean ambient space. It is given in terms of the microstructure function $\tilde{\varphi}_t$. The spatial velocity $\tilde{\boldsymbol{v}}$ of $\boldsymbol{p}$ is calculated by means of the time derivative of $\tilde{\varphi}_t$, i.e.,

$$\tilde{\boldsymbol{v}}(\boldsymbol{x},t) = \left(\left. \frac{\partial \tilde{\varphi}_t(\boldsymbol{X})}{\partial t} \right|_{\boldsymbol{X}=\text{const.}} \right) \circ {\varphi_t}^{-1}, \quad \boldsymbol{p} = \tilde{\varphi}_t(\boldsymbol{X}). \tag{3.43}$$

Regarding Eq. (2.28), the considered extensions to the energy balance, based on the vectorial DOF, are given by

$$\kappa = \frac{1}{2}\tilde{A}\tilde{\boldsymbol{v}} \cdot \tilde{\boldsymbol{v}}, \quad \beta = \tilde{\boldsymbol{b}} \cdot \tilde{\boldsymbol{v}}\,, \quad s = \tilde{\boldsymbol{t}} \cdot \tilde{\boldsymbol{v}}\,. \tag{3.44}$$

Here, $\tilde{\boldsymbol{b}}$ denotes a generalized micro-body force and $\tilde{\boldsymbol{t}}$ a generalized micro-traction. The micro-inertia is referred to as $\tilde{A}$. Thus, the extended energy balance Eq. (2.28) can be written in the form

$$\begin{aligned} \epsilon = \int_{\mathcal{V}_t} & (\dot{\rho} + \rho \operatorname{div}(\boldsymbol{v})) \left(e + \frac{1}{2}\boldsymbol{v} \cdot \boldsymbol{v} + \frac{1}{2}\tilde{A}\tilde{\boldsymbol{v}} \cdot \tilde{\boldsymbol{v}} \right) + \rho \dot{e} + \rho\left(\boldsymbol{a} - \boldsymbol{b}\right) \cdot \boldsymbol{v} \\ & + \rho\left(\tilde{A}\tilde{\boldsymbol{a}} - \tilde{\boldsymbol{b}}\right) \cdot \tilde{\boldsymbol{v}} + \frac{1}{2}\rho\dot{\tilde{A}}\tilde{\boldsymbol{v}} \cdot \tilde{\boldsymbol{v}} - \rho r \,\mathrm{d}v - \int_{\partial\mathcal{V}_t} \boldsymbol{t} \cdot \boldsymbol{v} + \tilde{\boldsymbol{t}} \cdot \tilde{\boldsymbol{v}} + h \,\mathrm{d}a = 0. \end{aligned} \tag{3.45}$$

The additional vectorial DOF $\boldsymbol{p}$ is assumed to transform objective concerning a change of observer, i.e.,

$$\boldsymbol{p}' = \boldsymbol{Q}(t)\boldsymbol{p} \tag{3.46}$$

holds true. Thus, $\boldsymbol{p}$ is unaffected if a pure translational transformation is considered. The implications for the spatial velocity fields are given in the first row of Tab. 3.2. The considered calculations are closely related to the discussion given in Marsden and Hughes (1994, pp. 145-149). As for the additional scalar-valued DOF, a pure translational transformation is considered first, and a pure rotational one subsequently.

3.2.2 Translational transformation

Since the same transformations are considered as discussed in Section 3.1.2, the additional DOFs are not affected by the rigid-body translation. Thus, the existence of the Cauchy stress, cf. Eq. (3.6) as well as the conservation of mass, cf. Eq. (3.10) and the balance of linear momentum, cf. Eq. (3.12) are obtained, respectively.

	$\boldsymbol{x}'$	$\boldsymbol{v}'$	$\boldsymbol{w}$	$\boldsymbol{p}'$	$\tilde{\boldsymbol{v}}'$	$\boldsymbol{z}$
Euclidean	$\boldsymbol{Q}\boldsymbol{x}+\boldsymbol{c}'$	$\boldsymbol{Q}\boldsymbol{v}+\boldsymbol{w}$	$\dot{\boldsymbol{Q}}\boldsymbol{x}+\dot{\boldsymbol{c}}'$	$\boldsymbol{Q}\boldsymbol{p}$	$\boldsymbol{Q}\tilde{\boldsymbol{v}}+\boldsymbol{z}$	$\dot{\boldsymbol{Q}}\boldsymbol{p}$
Translation	$\boldsymbol{x}+\boldsymbol{c}'$	$\boldsymbol{v}+\boldsymbol{w}$	$\dot{\boldsymbol{c}}'$	$\boldsymbol{p}$	$\tilde{\boldsymbol{v}}$	$\boldsymbol{0}$
Rotation	$\boldsymbol{Q}\boldsymbol{x}$	$\boldsymbol{Q}\boldsymbol{v}+\boldsymbol{w}$	$\dot{\boldsymbol{Q}}\boldsymbol{x}$	$\boldsymbol{Q}\boldsymbol{p}$	$\boldsymbol{Q}\tilde{\boldsymbol{v}}+\boldsymbol{z}$	$\dot{\boldsymbol{Q}}\boldsymbol{p}$

Table 3.2: Transformations of the material points and the additionally considered vectorial DOF in the context of a change of observer.

3.2.3 Rotational transformation

Accounting for the results in Eqs. (3.6), (3.10) and (3.12), a pure rotational transformation is considered subsequently, according to the transformation laws in the third row of Tab. 3.2. In line with Eq. (3.14), the traction forces are assumed to transform objectively with respect to a change of observer, reading

$$\boldsymbol{t}' = \boldsymbol{Q}\boldsymbol{t}, \quad \tilde{\boldsymbol{t}}' = \boldsymbol{Q}\tilde{\boldsymbol{t}}. \tag{3.47}$$

Similar to Eq. (3.15), the difference between body forces and accelerations is assumed to transform objectively

$$(\boldsymbol{a}' - \boldsymbol{b}') = \boldsymbol{Q}\,(\boldsymbol{a} - \boldsymbol{b})\,, \quad \left(\tilde{A}'\tilde{\boldsymbol{a}}' - \tilde{\boldsymbol{b}}'\right) = \boldsymbol{Q}\,(\tilde{A}\tilde{\boldsymbol{a}} - \tilde{\boldsymbol{b}})\,, \tag{3.48}$$

cf. Marsden and Hughes (1994). As in Section 3.1.3, all scalar quantities are invariant with respect to the considered transformations. It is assumed that $\boldsymbol{Q}(t_0) = \boldsymbol{I}$ holds true.

Existence of gradient stress tensor Evaluation of Eq. (4.16) yields

$$\begin{aligned}\Delta\epsilon_{t_0} = \int_{\mathcal{V}_t} &- \boldsymbol{\sigma} \cdot \operatorname{grad}(\boldsymbol{w}) + \rho\,(\tilde{A}\tilde{\boldsymbol{a}} - \tilde{\boldsymbol{b}}) \cdot \boldsymbol{z} + \rho\dot{\tilde{A}}\left(\tilde{\boldsymbol{v}} \cdot \boldsymbol{z} + \frac{1}{2}\boldsymbol{z} \cdot \boldsymbol{z}\right)\,\mathrm{d}v \\ &- \int_{\partial\mathcal{V}_t} \tilde{\boldsymbol{t}} \cdot \boldsymbol{z}\,\mathrm{d}a = 0. \end{aligned} \tag{3.49}$$

Application of Eq. (3.49) to an infinitesimal tetrahedron yields the existence of the gradient stress tensor $\tilde{\boldsymbol{\sigma}}$ given by

$$\tilde{\boldsymbol{\sigma}}\boldsymbol{n} = \tilde{\boldsymbol{t}}, \tag{3.50}$$

cf. Yavari and Marsden (2009, p. 9).

Conservation of micro-inertia Accounting for Eq. (3.50) and applying the divergence theorem, Eq. (3.49) can be formulated as

$$\begin{aligned}\Delta\epsilon_{t_0} = \int_{\mathcal{V}_t} &- \boldsymbol{\sigma} \cdot \operatorname{grad}(\boldsymbol{w}) + \rho\,(\tilde{A}\tilde{\boldsymbol{a}} - \tilde{\boldsymbol{b}}) \cdot \boldsymbol{z} + \rho\dot{\tilde{A}}\left(\tilde{\boldsymbol{v}} \cdot \boldsymbol{z} + \frac{1}{2}\boldsymbol{z} \cdot \boldsymbol{z}\right) \\ &- \operatorname{div}(\tilde{\boldsymbol{\sigma}}) \cdot \boldsymbol{z} - \tilde{\boldsymbol{\sigma}} \cdot \operatorname{grad}(\boldsymbol{z})\,\mathrm{d}v = 0. \end{aligned} \tag{3.51}$$

Moreover, substituting $\boldsymbol{w}$ and $\boldsymbol{z}$ according to the third row of Tab. 3.2 yields

$$\begin{aligned}\Delta\epsilon_{t_0} = \int_{\mathcal{V}_t} & -\boldsymbol{\sigma}\cdot\operatorname{grad}\left(\dot{\boldsymbol{Q}}\boldsymbol{x}\right) + \rho\left(\tilde{A}\tilde{\boldsymbol{a}} - \tilde{\boldsymbol{b}}\right)\cdot\left(\dot{\boldsymbol{Q}}\boldsymbol{p}\right) \\ & + \rho\dot{\tilde{A}}\left(\tilde{\boldsymbol{v}}\cdot\left(\dot{\boldsymbol{Q}}\boldsymbol{p}\right) + \frac{1}{2}\left(\dot{\boldsymbol{Q}}\boldsymbol{p}\right)\cdot\left(\dot{\boldsymbol{Q}}\boldsymbol{p}\right)\right) \\ & - \operatorname{div}\left(\tilde{\boldsymbol{\sigma}}\right)\cdot\left(\dot{\boldsymbol{Q}}\boldsymbol{p}\right) - \tilde{\boldsymbol{\sigma}}\cdot\operatorname{grad}\left(\dot{\boldsymbol{Q}}\boldsymbol{p}\right)\,\mathrm{d}v = 0. \end{aligned} \tag{3.52}$$

Manipulations of Eq. (3.52) lead to

$$\begin{aligned}\Delta\epsilon_{t_0} = \int_{\mathcal{V}_t} & -\boldsymbol{\sigma}\cdot\dot{\boldsymbol{Q}} + \rho\left(\tilde{A}\tilde{\boldsymbol{a}} - \tilde{\boldsymbol{b}}\right)\cdot\left(\dot{\boldsymbol{Q}}\boldsymbol{p}\right) \\ & + \rho\dot{\tilde{A}}\left(\tilde{\boldsymbol{v}}\cdot\left(\dot{\boldsymbol{Q}}\boldsymbol{p}\right) + \frac{1}{2}\left(\dot{\boldsymbol{Q}}^{\mathsf{T}}\dot{\boldsymbol{Q}}\right)\cdot\left(\boldsymbol{p}\otimes\boldsymbol{p}\right)\right) \\ & - \operatorname{div}\left(\tilde{\boldsymbol{\sigma}}\right)\cdot\left(\dot{\boldsymbol{Q}}\boldsymbol{p}\right) - \tilde{\boldsymbol{\sigma}}\cdot\left(\dot{\boldsymbol{Q}}\operatorname{grad}\left(\boldsymbol{p}\right)\right)\,\mathrm{d}v = 0. \end{aligned} \tag{3.53}$$

In general, a rotation tensor can be expressed by means of its rotation axis $\boldsymbol{n}$ and its rotation angle θ, cf. Spring (1986), reading

$$\boldsymbol{Q} = \boldsymbol{n}\otimes\boldsymbol{n} + \cos\theta\left(\boldsymbol{I} - \boldsymbol{n}\otimes\boldsymbol{n}\right) - \sin\theta\boldsymbol{\epsilon}[\boldsymbol{n}]. \tag{3.54}$$

If the rotation axis is considered as a constant unit vector, the time dependency of $\boldsymbol{Q}$ is due to the rotation angle $\theta = \theta(t)$. Thus, the time derivative of the rotation tensor is given by

$$\dot{\boldsymbol{Q}}(t) = -\left(\sin\theta\right)\dot{\theta}\boldsymbol{A} - \left(\cos\theta\right)\dot{\theta}\boldsymbol{B}, \tag{3.55}$$

with $\boldsymbol{A} = (\boldsymbol{I} - \boldsymbol{n} \otimes \boldsymbol{n})$ and $\boldsymbol{B} = \boldsymbol{\epsilon}[\boldsymbol{n}]$. Thus, the second derivative of Eq. (3.53) with respect to $\dot{\theta}$ reads

$$\begin{aligned} \frac{d^2 \Delta\epsilon_{t_0}}{d\dot{\theta}^2} &= \int_{\mathcal{V}_t} \frac{1}{2} \rho \dot{\tilde{A}} \frac{d^2 \left(\dot{\boldsymbol{Q}}^{\mathsf{T}} \dot{\boldsymbol{Q}}\right)}{d\dot{\theta}^2} \cdot (\boldsymbol{p} \otimes \boldsymbol{p}) \,\mathrm{d}v \\ &= \int_{\mathcal{V}_t} \rho \dot{\tilde{A}} \left((\sin\theta)^2 \boldsymbol{A}\boldsymbol{A} - (\cos\theta)^2 \boldsymbol{B}\boldsymbol{B}\right) \cdot (\boldsymbol{p} \otimes \boldsymbol{p}) \,\mathrm{d}v = 0. \end{aligned} \tag{3.56}$$

Localization of Eq. (3.56) yields

$$\rho \dot{\tilde{A}} \left((\sin\theta)^2 \boldsymbol{A}\boldsymbol{A} - (\cos\theta)^2 \boldsymbol{B}\boldsymbol{B}\right) \cdot (\boldsymbol{p} \otimes \boldsymbol{p}) = 0. \tag{3.57}$$

Since $\rho \neq 0$, $(\sin\theta)^2 \boldsymbol{A}\boldsymbol{A} - (\cos\theta)^2 \boldsymbol{B}\boldsymbol{B} \neq \boldsymbol{0}$ and $\boldsymbol{p} \otimes \boldsymbol{p} \neq \boldsymbol{0}$, the conservation of micro-inertia is obtained, reading

$$\dot{\tilde{A}} = 0. \tag{3.58}$$

Balance of angular momentum Accounting for conservation of micro-inertia as stated in Eq. (3.58), Eq. (3.53) can be formulated as

$$\begin{aligned} \Delta\epsilon_{t_0} = \int_{\mathcal{V}_t} \Big(&- \boldsymbol{\sigma} + \left(\rho \left(\tilde{A}\tilde{\boldsymbol{a}} - \tilde{\boldsymbol{b}}\right) - \operatorname{div}(\tilde{\boldsymbol{\sigma}})\right) \otimes \boldsymbol{p} \\ &- \tilde{\boldsymbol{\sigma}} (\operatorname{grad}(\boldsymbol{p}))^{\mathsf{T}} \Big) \cdot \dot{\boldsymbol{Q}} \,\mathrm{d}v = 0. \end{aligned} \tag{3.59}$$

Since $\dot{\boldsymbol{Q}}(t_0) \in \mathit{Skw}$, the localization of Eq. (3.59) yields the modified balance of angular momentum

$$-\boldsymbol{\sigma} + \left(\rho \left(\tilde{A}\tilde{\boldsymbol{a}} - \tilde{\boldsymbol{b}}\right) - \operatorname{div}(\tilde{\boldsymbol{\sigma}})\right) \otimes \boldsymbol{p} - \tilde{\boldsymbol{\sigma}} (\operatorname{grad}(\boldsymbol{p}))^{\mathsf{T}} \in \mathit{Sym}. \tag{3.60}$$

Consequently, the Cauchy stress $\boldsymbol{\sigma}$ is not symmetric as in Eq. (2.15).

Existence of heat flux vector Accounting for the results in Eqs. (3.6), (3.10), (3.12), (3.50) and (3.58), Eq. (3.45) can be written as

$$\int_{\mathcal{V}_t} \rho\dot{e} - \boldsymbol{\sigma}\cdot\operatorname{grad}(\boldsymbol{v}) + \left(\rho\left(\tilde{A}\tilde{\boldsymbol{a}} - \tilde{\boldsymbol{b}}\right) - \operatorname{div}(\tilde{\boldsymbol{\sigma}})\right)\cdot\tilde{\boldsymbol{v}} \\ - \tilde{\boldsymbol{\sigma}}\cdot\operatorname{grad}(\tilde{\boldsymbol{v}}) - \rho r\,\mathrm{d}v - \int_{\partial\mathcal{V}_t} h\,\mathrm{d}a = 0. \tag{3.61}$$

Application of Eq. (3.61) to an infinitesimal tetrahedron yields the existence of the heat flux vector $\boldsymbol{q}$ given by

$$\boldsymbol{q}\cdot\boldsymbol{n} = -h, \tag{3.62}$$

cf. Marsden and Hughes (1994, p. 148).

Balance of internal energy Under consideration of all previously discussed results, localization of Eq. (3.45) yields the local form of the balance of internal energy, reading

$$\rho\dot{e} - \boldsymbol{\sigma}\cdot\operatorname{grad}(\boldsymbol{v}) + \left(\rho\left(\tilde{A}\tilde{\boldsymbol{a}} - \tilde{\boldsymbol{b}}\right) - \operatorname{div}(\tilde{\boldsymbol{\sigma}})\right)\cdot\tilde{\boldsymbol{v}} \\ - \rho r - \tilde{\boldsymbol{\sigma}}\cdot\operatorname{grad}(\tilde{\boldsymbol{v}}) + \operatorname{div}(\boldsymbol{q}) = 0. \tag{3.63}$$

3.3 Interim Conclusions

Subsequently, the results of the current chapter are briefly summarized and concluding remarks are given.

- An additional balance equation for the supplementary introduced degree of freedom cannot be obtained by invariance considerations.
- The existence of a conservation law for micro-inertia cannot be obtained for a scalar-valued DOF, but for a vector-valued DOF.
- The unambiguous existence of the Cauchy stress tensor, the gradient stress vector and the heat flux vector can be shown. This is in direct

contrast to a corresponding extended principle of virtual power, which assumes the existence of stress quantities from the outset.

- The conservation of mass and the balance of linear momentum are not affected by contributions associated with the additional DOF.
- In addition, the Cauchy stress tensor remains symmetric only for a scalar-valued DOF.
- A thermodynamically consistent nonlocal flow rule for the scalar-valued DOF is obtained by the exploitation of the Clausius-Duhem inequality.
- The so-called micro-force balance results from partitioning of the nonlocal flow rule.
- Regarding an isothermal behavior with a homogeneous temperature distribution and neglecting effects due to heat supply, the nonlocal flow rule for the additional DOF reduces to a PDE.
- In the context of an additional scalar DOF, the equivalence between the extended energy balance and an extended principle of virtual power is shown. Thereby the micro-inertia and micro-body forces are neglected. It is outlined, that the notion of balance is misleading with respect to the micro-force balance.

Chapter 4

On Interface Conditions on a Material Singular Surface[1]

Motivation The presence of grain boundaries (GBs) significantly influences the overall mechanical behavior of materials with an underlying crystalline microstructure. They act as an obstacle against the movement of dislocations and, thus, significantly contribute to size effects as for example the Hall-Petch effect. Hence, the thermodynamically consistent modeling of the behavior of the plastic slip at a GB is of utmost interest. To this end, balance equations at a GB are derived from an extended energy balance by means of invariance considerations. The GB is considered as a material singular surface with own internal and kinetic energy as well as energy supply. Consequently, the balances at the grain boundary depend on its mean curvature. The implications of the mean curvature of the GB on the balance equations are briefly addressed. The framework presented is applied to a small strain slip gradient crystal plasticity theory, regarding single slip. Accounting for the derived balance equations, thermodynamically consistent flow rules for the plastic slip at the GB are obtained by exploitation of the Coleman-Noll procedure. In this context, a classification of flow rules for the plastic slip at the GB is provided. The equivalence between the illustrated theory and an extended principle of virtual power is shown.

[1] The content of this chapter is taken directly from the article Prahs and Böhlke (2019a). Additional content concerning the connection to an extended principle of virtual power is considered. Minor linguistic changes and abbreviations have been made.

4.1 Balance Equations and Dissipation Inequality

4.1.1 Extended energy balance

A material volume divided by a material singular surface Figure 4.1 illustrates a material volume $\mathcal{V}$ separated by a material singular surface $\mathcal{S}$ into two subvolumes $\mathcal{V}^+$ and $\mathcal{V}^-$. The subvolumes $\mathcal{V}^+$ and $\mathcal{V}^-$ are bounded by the surfaces $\mathcal{F}^+$ and $\mathcal{F}^-$ towards the surrounding and separated by the singular surface $\mathcal{S}$ from each other. While $\boldsymbol{n}_{\mathcal{V}}$ denotes the outer normal vector of the surfaces $\mathcal{F}^+$ and $\mathcal{F}^-$, $\boldsymbol{n}_{\mathcal{S}}$ denotes the normal vector of the singular surface. Here, $\boldsymbol{n}_{\mathcal{S}}$ points from $\mathcal{V}^-$ to $\mathcal{V}^+$. As result of the so called pill-box theorem, cf. Müller (1985), $\boldsymbol{n}_{\mathcal{S}} = \boldsymbol{n}_{\mathcal{V}}^+ = -\boldsymbol{n}_{\mathcal{V}}^-$ holds true. Moreover, the outer normal vector of the boundary $\partial\mathcal{S}$ of the singular surface $\mathcal{S}$ is denoted as $\boldsymbol{m}$. Regarding an arbitrary quantity a, its jump at a singular surface is defined as $\{\!\!\{a\}\!\!\} = a^+ - a^-$, with a^+ and a^- denoting the right- and left-hand limit of a, respectively, cf. Truesdell and Toupin (1960, p. 492).

Deformation and microstructure function The position of a material point is denoted as $\boldsymbol{x}$ regarding the current configuration. It can be described in dependency of the position of the material point $\boldsymbol{X}$ in the reference configuration by means of the deformation function φ_t. Thus, the spatial velocity field $\boldsymbol{v}(\boldsymbol{x}, t)$ can also be expressed in terms of the deformation function, reading

$$\boldsymbol{v}(\boldsymbol{x}, t) = \left(\left. \frac{\partial \varphi_t(\boldsymbol{X})}{\partial t} \right|_{\boldsymbol{X}=\text{const.}} \right) \circ {\varphi_t}^{-1}, \quad \boldsymbol{x} = \varphi_t(\boldsymbol{X}). \tag{4.1}$$

Here, the symbol $\circ$ is used to describe the composition between two maps. A generic additional scalar-valued DOF is introduced that is related to the evolution of the microstructure. In a subsequent chapter, the framework presented is applied to a slip gradient crystal plasticity theory.

In this context, the additional DOF is considered as plastic slip. Similar to Eq. (4.1), the additional DOF γ as well as its rate $\dot{\gamma}$ can be described using a microstructure function $\tilde{\varphi}_t$, reading

$$\dot{\gamma}(\boldsymbol{x}, t) = \left(\left. \frac{\partial \tilde{\varphi}_t(\boldsymbol{X})}{\partial t} \right|_{\boldsymbol{X}=\text{const.}} \right) \circ \boldsymbol{\varphi}_t{}^{-1}, \quad \gamma = \tilde{\varphi}_t(\boldsymbol{X}). \tag{4.2}$$

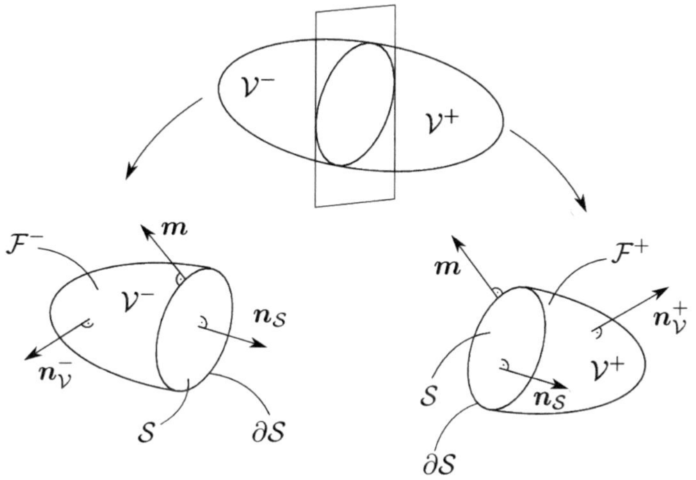

Figure 4.1: Material volume, separated by a material singular surface $\mathcal{S}$, cf. also Müller (1985).

Energy balance Regarding the current configuration $\mathcal{C}$ of a material volume, an extended balance of total energy is given by

$$
\begin{aligned}
&\frac{\mathrm{d}}{\mathrm{d}t} \int\limits_{\mathcal{V}^+\cup\mathcal{V}^-} \rho\left(e+\frac{1}{2}\boldsymbol{v}\cdot\boldsymbol{v}+\frac{1}{2}\tilde{A}\dot{\gamma}^2\right)\,\mathrm{d}v+\frac{\mathrm{d}}{\mathrm{d}t}\int\limits_{\mathcal{S}}\rho_{\mathcal{S}}\left(e_{\mathcal{S}}+\frac{1}{2}\boldsymbol{v}^{\mathcal{S}}\cdot\boldsymbol{v}^{\mathcal{S}}\right)\,\mathrm{d}a \\
&-\int\limits_{\mathcal{V}^+\cup\mathcal{V}^-}\rho\left(\boldsymbol{b}\cdot\boldsymbol{v}+\tilde{b}\dot{\gamma}+r\right)\,\mathrm{d}v-\int\limits_{\mathcal{F}^+\cup\mathcal{F}^-}\boldsymbol{t}\cdot\boldsymbol{v}+\tilde{t}\dot{\gamma}+h\,\mathrm{d}a \\
&-\int\limits_{\mathcal{S}}\rho_{\mathcal{S}}\left(\boldsymbol{b}^{\mathcal{S}}\cdot\boldsymbol{v}^{\mathcal{S}}+r_{\mathcal{S}}\right)\,\mathrm{d}a=0. \qquad (4.3)
\end{aligned}
$$

The contributions $\tilde{A}\dot{\gamma}^2/2$, $\tilde{b}\dot{\gamma}$ and $\tilde{t}\dot{\gamma}$ are associated with the additional, scalar-valued DOF γ, cf. Prahs and Böhlke (2019b), and, thus, supplement the balance of total energy of a classical continuum, cf. Müller (1985). In this context, $\tilde{A}$ denotes the micro-inertia, $\tilde{b}$ and $\tilde{t}$ the generalized body and traction forces associated with γ. For brevity only one additional degree of freedom is considered, here. The mass densities with respect to $\mathcal{V}^+\cup\mathcal{V}^-$ and $\mathcal{S}$ are referred to as ρ and $\rho_{\mathcal{S}}$, respectively. Here, e and $e_{\mathcal{S}}$ denote the specific internal energy with respect to the bulk material and the singular surface, respectively. The spatial velocity field of the body is denoted by $\boldsymbol{v}$ and the velocity field of the singular surface by $\boldsymbol{v}^{\mathcal{S}}$. Here, a material singular surface is considered, which is interpreted as a GB in the context of a crystal plasticity framework. Thus, $\boldsymbol{v}^{\mathcal{S}}=\boldsymbol{v}$ and $\{\!\{\boldsymbol{v}\}\!\}=\boldsymbol{0}$ hold true, cf. (Cermelli et al., 2005, pp. 345, 346). The velocity of the surface consists of a contribution $\boldsymbol{v}_{||}$ parallel and a contribution $\boldsymbol{v}_{\perp}$ perpendicular to $\mathcal{S}$. In contrast to a non-material surface, for a material one the tangential velocity $\boldsymbol{v}_{||}$ can be uniquely defined. Moreover, the normal component of the velocity of the singular surface is given by $v_{\perp}^{\mathcal{S}}\equiv v_{\perp}=\boldsymbol{v}\cdot\boldsymbol{n}_{\mathcal{S}}$. Concerning the bulk material and the singular surface, $\boldsymbol{b}$ and $\boldsymbol{b}^{\mathcal{S}}$ denote the specific body force, respectively. The surface traction is referred to as $\boldsymbol{t}$. While r denotes the specific energy supply regarding the bulk material, $r_{\mathcal{S}}$ denotes the specific energy supply concerning the singular surface. The heat flux is referred to as h.

4.1.2 Mathematical preliminaries

Divergence theorem In the presence of a singular surface the divergence theorem is given by

$$\int\limits_{\mathcal{F}^+\cup\mathcal{F}^-} \boldsymbol{a}\cdot\boldsymbol{n}\,\mathrm{d}a = \int\limits_{\mathcal{V}^+\cup\mathcal{V}^-} \operatorname{div}(\boldsymbol{a})\,\mathrm{d}v + \int\limits_{\mathcal{S}} \{\!\![\boldsymbol{a}]\!\!\}\cdot\boldsymbol{n}_{\mathcal{S}}\,\mathrm{d}a, \tag{4.4}$$

where $\{\!\![\boldsymbol{a}]\!\!\}\cdot\boldsymbol{n}_{\mathcal{S}} = \{\!\![\boldsymbol{a}\cdot\boldsymbol{n}_{\mathcal{S}}]\!\!\}$ holds true, cf. Gurtin et al. (2010, p. 215). Here, $\{\!\![\boldsymbol{a}]\!\!\}$ denotes the jump of an arbitrary vector field $\boldsymbol{a}$ at the singular surface.

Transport theorem for a material volume divided by a singular surface In order to derive the local form of the energy balance according to Eq. (4.3), transport theorems for the volume and the singular surface have to be applied. In the context of a material singular surface, Reynolds' transport theorem reads

$$\frac{\mathrm{d}}{\mathrm{d}t}\int\limits_{\mathcal{V}^+\cup\mathcal{V}^-} \Psi_{\mathcal{V}}\,\mathrm{d}v = \int\limits_{\mathcal{V}^+\cup\mathcal{V}^-} \frac{\partial\Psi_{\mathcal{V}}}{\partial t}\,\mathrm{d}v + \int\limits_{\mathcal{F}^+\cup\mathcal{F}^-} \Psi_{\mathcal{V}}\,\boldsymbol{v}\cdot\boldsymbol{n}\,\mathrm{d}a - \int\limits_{\mathcal{S}} \{\!\![\Psi_{\mathcal{V}}]\!\!\}\,v_{\perp}\,\mathrm{d}a, \tag{4.5}$$

cf. Truesdell and Toupin (1960, p. 526). Here, $\Psi_{\mathcal{V}}$ denotes a bulk density. Application of the divergence theorem according to Eq. (4.4) yields a formulation without a jump

$$\frac{\mathrm{d}}{\mathrm{d}t}\int\limits_{\mathcal{V}^+\cup\mathcal{V}^-} \Psi_{\mathcal{V}}\,\mathrm{d}v = \int\limits_{\mathcal{V}^+\cup\mathcal{V}^-} \dot{\Psi}_{\mathcal{V}} + \Psi_{\mathcal{V}}\operatorname{div}(\boldsymbol{v})\,\mathrm{d}v, \tag{4.6}$$

cf. Cermelli et al. (2005).

Transport theorem for a material surface Regarding a singular surface, the differential geometry of the surface has to be taken into account, cf. Moeckel (1975). The transport theorem for a material surface is given by

$$\frac{\mathrm{d}}{\mathrm{d}t}\int\limits_{\mathcal{S}} \Psi_{\mathcal{S}}\,\mathrm{d}a = \int\limits_{\mathcal{S}} \dot{\Psi}_{\mathcal{S}} + \Psi_{\mathcal{S}}\left\{\mathrm{div}_{\mathcal{S}}\left(\boldsymbol{v}_{||}\right) - v_{\perp}2K_{\mathrm{m}}\right\}\,\mathrm{d}a, \tag{4.7}$$

cf. Müller (1985, p. 50) and Cermelli et al. (2005, p. 346). Here,

$$\dot{\Psi}_{\mathcal{S}} := \frac{\partial \Psi_{\mathcal{S}}}{\partial t} \tag{4.8}$$

refers to the corresponding material time derivative of $\Psi_{\mathcal{S}}$. Since a material singular surface is considered, the surface coordinates are material coordinates and, thus, regarded as constant with respect to time. The divergence $\mathrm{div}_{\mathcal{S}}(\cdot)$ is a planar operator, cf. Slattery et al. (2007, p. 647) for a detailed discussion on surface differential operators. The mean curvature K_{m} of the surface is correlated to the trace of its curvature tensor $\boldsymbol{K}$ as $K_{\mathrm{m}} = \mathrm{tr}(\boldsymbol{K})/2$, cf. Itskov (2015, p. 82) or Müller (1985, p. 44). Moreover, K_{m} can be expressed in terms of the surface divergence of the outer normal vector of the singular surface, reading

$$2K_{\mathrm{m}} = -\mathrm{div}_{\mathcal{S}}\left(\boldsymbol{n}_{\mathcal{S}}\right), \tag{4.9}$$

cf. Cermelli et al. (2005, p. 341). Hence, it is possible to write

$$\begin{aligned}
\mathrm{div}_{\mathcal{S}}\left(\boldsymbol{v}_{||}\right) - v_{\perp}2K_{\mathrm{m}} &= \mathrm{div}_{\mathcal{S}}\left(\boldsymbol{v}_{||}\right) + \left(\boldsymbol{v}\cdot\boldsymbol{n}_{\mathcal{S}}\right)\mathrm{div}_{\mathcal{S}}\left(\boldsymbol{n}_{\mathcal{S}}\right) \\
&= \mathrm{div}_{\mathcal{S}}\left(\boldsymbol{v}_{||}\right) + \mathrm{div}_{\mathcal{S}}\left(\left(\boldsymbol{v}\cdot\boldsymbol{n}_{\mathcal{S}}\right)\boldsymbol{n}_{\mathcal{S}}\right) - \boldsymbol{n}_{\mathcal{S}}\cdot\mathrm{grad}_{\mathcal{S}}\left(\boldsymbol{v}\cdot\boldsymbol{n}_{\mathcal{S}}\right) \\
&= \mathrm{div}_{\mathcal{S}}\left(\boldsymbol{v}_{||}\right) + \mathrm{div}_{\mathcal{S}}\left(\boldsymbol{v}_{\perp}\right) \\
&= \mathrm{div}_{\mathcal{S}}\left(\boldsymbol{v}\right),
\end{aligned} \tag{4.10}$$

cf. Cermelli et al. (2005). The surface gradient $\mathrm{grad}_{\mathcal{S}}\left(\boldsymbol{v}\cdot\boldsymbol{n}_{\mathcal{S}}\right)$ is a vector field tangential to the surface. Consequently, the scalar product

$\boldsymbol{n}_{\mathcal{S}} \cdot \operatorname{grad}_{\mathcal{S}} (\boldsymbol{v} \cdot \boldsymbol{n}_{\mathcal{S}})$ vanishes. Thus, Eq. (4.7) can be expressed as

$$\frac{\mathrm{d}}{\mathrm{d}t} \int_{\mathcal{S}} \Psi_{\mathcal{S}} \, \mathrm{d}a = \int_{\mathcal{S}} \dot{\Psi}_{\mathcal{S}} + \Psi_{\mathcal{S}} \operatorname{div}_{\mathcal{S}} (\boldsymbol{v}) \, \mathrm{d}a. \tag{4.11}$$

In contrast to Eq. (4.7), this specific form of the transport theorem does not explicitly contain a contribution that is related to the mean curvature K_{m}.

4.1.3 Invariance considerations of an extended energy balance

Balances in singular points Balance equations in singular points are commonly provided by means of the localization of a master balance equation that accounts for a singular surface, cf. Müller (1985, p. 52). In this context, specific balance equations are obtained by substituting the respective terms of the master balance equation. Thus, each balance equation is treated separately. In contrast, here, the balance equations in singular points are derived by means of the invariance considerations presented for regular points. Subsequently, the corresponding calculations are carried out in detail to ensure that the work at hand is self-contained. Accounting for the results according to Eqs. (3.6), (3.10), (3.12), (3.18), (3.22) and (3.23), the transport theorems according to Eq. (4.6) and Eq. (4.11) as well as the divergence theorem according to Eq. (4.4) are applied to Eq. (4.3). Motivated by the pill-box theorem, cf. Müller (1985), only singular points are considered. Thus, localisation yields

$$\begin{aligned} \epsilon = {} & (\dot{\rho}_{\mathcal{S}} + \rho_{\mathcal{S}} \operatorname{div}_{\mathcal{S}} (\boldsymbol{v})) \left(e_{\mathcal{S}} + \frac{1}{2} \boldsymbol{v} \cdot \boldsymbol{v} \right) + \rho_{\mathcal{S}} \dot{e}_{\mathcal{S}} \\ & + \left(\rho_{\mathcal{S}} \boldsymbol{a} - \rho_{\mathcal{S}} \boldsymbol{b}^{\mathcal{S}} - [\![\boldsymbol{t}]\!] \right) \cdot \boldsymbol{v} - \rho_{\mathcal{S}} r_{\mathcal{S}} + [\![\boldsymbol{q}]\!] \cdot \boldsymbol{n}_{\mathcal{S}} - [\![\boldsymbol{\xi} \dot{\gamma}]\!] \cdot \boldsymbol{n}_{\mathcal{S}} = 0. \end{aligned} \tag{4.12}$$

Invariance of Eq. (4.12) with respect to a change of observer according to Eq. (2.13) is assumed. In line with Eq. (3.15), the transformation

$$\left(\boldsymbol{a}' - \boldsymbol{b}^{\mathcal{S}\prime}\right) = \boldsymbol{Q}\left(\boldsymbol{a} - \boldsymbol{b}^{\mathcal{S}}\right) \tag{4.13}$$

is assumed. Regarding Eq. (4.3), the generalized traction force $\tilde{t}$ and the heat flux h are invariant with respect to a change of observer according to Eq. (2.13). Both, $\tilde{t}$ and h are linear in $\boldsymbol{n}$ according to Eq. (3.22). Consequently, the heat flux vector $\boldsymbol{q}$ as well as the gradient stress vector $\boldsymbol{\xi}$ transform objectively regarding a change of observer, i.e.,

$$\boldsymbol{q}' = \boldsymbol{Q}\boldsymbol{q}, \quad \boldsymbol{\xi}' = \boldsymbol{Q}\boldsymbol{\xi}. \tag{4.14}$$

Moreover,

$$\mathrm{div}'_{\mathcal{S}}\left(\boldsymbol{v}'\right) = \mathrm{div}_{\mathcal{S}}\left(\boldsymbol{v}\right) \tag{4.15}$$

holds true. All scalar quantities as well as their rates are invariant with respect to the Euclidean transformation. The energy balance in singular points described by the observer of the vector space $\mathcal{W}$ is referred to as ϵ, cf. Eq. (4.12). Consequently, the energy balance in singular points described by the observer of $\mathcal{W}'$ is designated by ϵ'. To obtain the field equations, the difference

$$\Delta\epsilon_{t_0} = \left(\epsilon' - \epsilon\right)\big|_{t=t_0} \tag{4.16}$$

is introduced and evaluated. Regarding $t = t_0$, $\boldsymbol{Q}(t_0) = \boldsymbol{I}$ and $\dot{\boldsymbol{Q}}(t_0) \in Skw$ holds true, cf. Marsden and Hughes (1994). Subsequently, a pure translational transformation is considered according to Sec. 3.1.2. Thus, Eq. (4.16) yields

$$\Delta\epsilon_{t_0} = (\dot{\rho}_{\mathcal{S}} + \rho_{\mathcal{S}} \operatorname{div}_{\mathcal{S}}(\boldsymbol{v})) \left(\boldsymbol{v} \cdot \boldsymbol{w} + \frac{1}{2}\boldsymbol{w} \cdot \boldsymbol{w}\right) + \left(\rho_{\mathcal{S}}\boldsymbol{a} - \rho_{\mathcal{S}}\boldsymbol{b}^{\mathcal{S}} - \{\!\!\{\boldsymbol{t}\}\!\!\}\right) \cdot \boldsymbol{w} = 0. \tag{4.17}$$

Since the vector field $\boldsymbol{w}$ is arbitrary, $\boldsymbol{w} = \lambda\boldsymbol{u}$ is considered with a constant unit vector $\boldsymbol{u}$. Thus, Eq. (4.17) reads

$$\Delta\epsilon_{t_0} = (\dot{\rho}_{\mathcal{S}} + \rho_{\mathcal{S}} \operatorname{div}_{\mathcal{S}}(\boldsymbol{v})) \left(\lambda\boldsymbol{v} \cdot \boldsymbol{u} + \frac{1}{2}\lambda^2\boldsymbol{u} \cdot \boldsymbol{u}\right) + \lambda\left(\rho_{\mathcal{S}}\boldsymbol{a} - \rho_{\mathcal{S}}\boldsymbol{b}^{\mathcal{S}} - \{\!\!\{\boldsymbol{t}\}\!\!\}\right) \cdot \boldsymbol{u} = 0. \tag{4.18}$$

Differentiating Eq. (4.18) twice with respect to λ yields

$$\frac{d^2\Delta\epsilon_{t_0}}{d\lambda^2} = (\dot{\rho}_{\mathcal{S}} + \rho_{\mathcal{S}} \operatorname{div}_{\mathcal{S}}(\boldsymbol{v}))(\boldsymbol{u} \cdot \boldsymbol{u}) = 0. \tag{4.19}$$

Thus, the conservation of the mass of the material singular surface is obtained, reading

$$\dot{\rho}_{\mathcal{S}} + \rho_{\mathcal{S}} \operatorname{div}_{\mathcal{S}}(\boldsymbol{v}) = 0. \tag{4.20}$$

Taking into acount the identity given in Eq. (4.10), the balance of mass according to Eq. (4.20) can alternatively be written as

$$\dot{\rho}_{\mathcal{S}} + \rho_{\mathcal{S}} \left(\operatorname{div}_{\mathcal{S}}(\boldsymbol{v}_{||}) - v_{\perp}2K_{\mathrm{m}}\right) = 0. \tag{4.21}$$

Accounting for Eq. (4.20), the balance of linear momentum in singular points is obtained from Eq. (4.17) as

$$\rho_{\mathcal{S}}\boldsymbol{a} - \rho_{\mathcal{S}}\boldsymbol{b}^{\mathcal{S}} - \{\!\!\{\boldsymbol{t}\}\!\!\} = \boldsymbol{0}. \tag{4.22}$$

Finally, taking into account Eqs. (4.20) and (4.22) with respect to Eq. (4.12), the balance of internal energy for singular points is given by

$$\rho_S \dot{e}_S - \rho_S r_S + \{\!\!\{\boldsymbol{q}\}\!\!\} \cdot \boldsymbol{n}_S - \{\!\!\{\boldsymbol{\xi}\dot{\gamma}\}\!\!\} \cdot \boldsymbol{n}_S = 0. \tag{4.23}$$

The balance of mass in singular points, cf. Eq. (4.21), explicitly depends on the mean curvature K_{m}. However, the balance of linear momentum and internal energy in Eqs. (4.22) and (4.23) are affected by the mean curvature only via the balance of mass from Eq. (4.21).

4.1.4 Dissipation inequality

Entropy balance The global form of the entropy balance is given in its standard form, reading

$$\begin{aligned} &\frac{\mathrm{d}}{\mathrm{d}t}\left(\int\limits_{\mathcal{V}^+\cup\mathcal{V}^-} \rho\eta \,\mathrm{d}v + \int\limits_{\mathcal{S}} \rho_S \eta_S \,\mathrm{d}a\right) \\ &\quad = -\int\limits_{\mathcal{F}^+\cup\mathcal{F}^-} \boldsymbol{\phi}^{\eta}_{\mathcal{V}} \cdot \boldsymbol{n}_{\mathcal{V}} \,\mathrm{d}a + \int\limits_{\mathcal{V}^+\cup\mathcal{V}^-} \rho\, p^{\eta}_{\mathcal{V}} + s^{\eta}_{\mathcal{V}} \,\mathrm{d}v + \int\limits_{\mathcal{S}} \rho_S\, p^{\eta}_{S} + s^{\eta}_{S} \,da, \end{aligned} \tag{4.24}$$

cf. Müller (1985). While η denotes the specific entropy considering the bulk material, η_S indicates the specific entropy with respect to the singular surface. As common in classical thermodynamics, the entropy flux $\phi^{\eta}_{\mathcal{V}}$ across the boundary $\mathcal{F}^+ \cup \mathcal{F}^-$ is assumed to be given by the ratio of the heat flux $\boldsymbol{q}$ and the temperature θ, cf. , e.g., Coleman and Noll (1963). The specific entropy production is denoted as $p^{\eta}_{\mathcal{V}}$ and p^{η}_{S} regarding bulk material and the singular surface, respectively. In regular points, the entropy supply is assumed to be $s^{\eta}_{\mathcal{V}} = \rho r/\theta$. Accordingly, the entropy supply is assumed as $s^{\eta}_{S} = \rho_S r_S/\theta$ in singular points. Accounting for the assumptions concerning the entropy flux and supply, the dissipation inequality is also often referred to as Clausius-Duhem inequality, cf. Coleman and Gurtin (1967); Coleman and Noll (1963).

Clausius-Duhem inequality inequality in singular points On the singular surface, the dissipation is defined as $\delta_{\mathcal{S}} := p^{\eta}_{\mathcal{S}}\theta$. In analogy to the derivation for regular points, the relation between the specific internal energy $e_{\mathcal{S}}$, the specific free energy $\psi_{\mathcal{S}}$ and the specific entropy $\eta_{\mathcal{S}}$ on $\mathcal{S}$ is given by $\psi_{\mathcal{S}} = e_{\mathcal{S}} - \theta\eta_{\mathcal{S}}$. Moreover, the continuity of the temperature across the material singular surface is assumed, i.e., $[\![\theta]\!] = 0$, cf. Müller (1985, p. 11) and Struchtrup (2008, p. 494). Furthermore, taking into account Eq. (4.23), the Clausius-Duhem inequality in singular points reads

$$\rho_{\mathcal{S}}\delta_{\mathcal{S}} = -\rho_{\mathcal{S}}\dot{\psi}_{\mathcal{S}} - \rho_{\mathcal{S}}\dot{\theta}\eta_{\mathcal{S}} + [\![\boldsymbol{\xi}\dot{\gamma}]\!] \cdot \boldsymbol{n}_{\mathcal{S}} \geq 0. \tag{4.25}$$

This dissipation inequality poses restrictions on the constitutive equations that describe the material behavior, cf. Triani and Cimmelli (2012). In line with Eq. (4.23), the Clausius-Duhem inequality according to Eq. (4.25) is affected only implicitly by the mean curvature K_{m} in terms of the balance of mass according to Eq. (4.21). For brevity, the jump in Eq. (4.25) can be written as

$$\begin{aligned} [\![\boldsymbol{\xi}\dot{\gamma}]\!] \cdot \boldsymbol{n}_{\mathcal{S}} &= \xi_{\perp}^{+}\dot{\gamma}^{+} + \xi_{\perp}^{-}\dot{\gamma}^{-}, \\ \xi_{\perp}^{+} = \boldsymbol{\xi}^{+}\big|_{\mathcal{S}} \cdot \boldsymbol{n}_{\mathcal{V}}^{+}, &\quad \xi_{\perp}^{-} = \boldsymbol{\xi}^{-}\big|_{\mathcal{S}} \cdot \boldsymbol{n}_{\mathcal{V}}^{-}, \end{aligned} \tag{4.26}$$

with $\boldsymbol{n}_{\mathcal{S}} = \boldsymbol{n}_{\mathcal{V}}^{+} = -\boldsymbol{n}_{\mathcal{V}}^{-}$.

4.1.5 Potential relations and boundary conditions at the singular surface

Potential relations and reduced dissipation inequality The specific surface free energy $\psi_{\mathcal{S}}$ is assumed to consist of a contribution $\psi_{\mathcal{S}}^{\gamma}$ depending on the plastic slips of two adjacent grains and a contribution $\psi_{\mathcal{S}}^{\theta}$ depending on the absolute temperature. Therefore, the specific surface free energy can be written as

$$\psi_S\left(\gamma^+,\gamma^-,\theta\right) = \psi_S^\gamma\left(\gamma^+,\gamma^-\right) + \psi_S^\theta\left(\theta\right). \tag{4.27}$$

Thus, the Clausius-Duhem inequality in singular points, cf. Eq. (4.25), reads

$$\begin{aligned}\rho_S\delta_S = &\left(\xi_\perp^+ - \rho_S\frac{\partial\psi_S^\gamma}{\partial\gamma^+}\right)\dot{\gamma}^+ + \left(\xi_\perp^- - \rho_S\frac{\partial\psi_S^\gamma}{\partial\gamma^-}\right)\dot{\gamma}^- \\ &- \rho_S\left(\eta_S + \frac{\partial\psi_S^\theta}{\partial\theta}\right)\dot{\theta} \geq 0.\end{aligned} \tag{4.28}$$

Similar to regular points, the Coleman-Noll procedure is applied, cf. Triani and Cimmelli (2012). The potential relation

$$\eta_S = -\frac{\partial\psi_S^\theta}{\partial\theta} \tag{4.29}$$

for the entropy is obtained, leading to the reduced dissipation inequality in singular points

$$\left(\xi_\perp^+ - \rho_S\frac{\partial\psi_S^\gamma}{\partial\gamma^+}\right)\dot{\gamma}^+ + \left(\xi_\perp^- - \rho_S\frac{\partial\psi_S^\gamma}{\partial\gamma^-}\right)\dot{\gamma}^- \geq 0. \tag{4.30}$$

In general, the specific surface free energy ψ_S is considered inversely proportional to ρ_S. Thus, the mean curvature does not have to be taken into account for the constitutive modeling of the plastic slip at the GB. Subsequently admissible flow rules for γ^+ and γ^- are presented that fulfill the reduced dissipation inequality according to Eq. (4.30). If the rates $\dot{\gamma}^+$ and $\dot{\gamma}^-$ are considered independent of each other, the following special case of the rate equations for the slips can be discussed .

Rate-dependent, nonlinear A rate-dependent flow rule for modeling a viscoplastic behavior is given in terms of a power-law, reading

$$\dot{\gamma}^{\pm} = \dot{\gamma}_0^{\pm} \left\langle \frac{\xi_{\perp}^{\pm} - \rho_S \partial \psi_S^{\gamma} / \partial \gamma^{\pm}}{\tau_D^{\pm}} \right\rangle^{m} \operatorname{sgn}\left(\xi_{\perp}^{\pm} - \rho_S \partial \psi_S^{\gamma} / \partial \gamma^{\pm}\right). \tag{4.31}$$

The strain rate sensitivity is referred to as m and the respective drag stress as $\tau_D^{\pm}$. Moreover, $\dot{\gamma}_0^{\pm}$ denote referential shear rates in the respective grain, and $\xi_{\perp}^{\pm}$ is defined according to Eq. (4.26). The Macaulay brackets are denoted by $\langle \cdot \rangle$. Thus, the expression $\langle a \rangle$ takes the value a, if $a \geq 0$ and vanishes otherwise.

Linear irreversible thermodynamics In this context, choosing the rate of the plastic slips $\dot{\gamma}^{+}$ and $\dot{\gamma}^{-}$ proportional to their corresponding prefactor in Eq. (4.30) guarantees the positivity of the reduced dissipation inequality. Thus, the flow rule according to linear irreversible thermodynamics constitutes the special case of Eq. (4.31) for $m = 1$, reading

$$\begin{aligned} \dot{\gamma}^{\pm} &= \dot{\gamma}_0^{\pm} \left\langle \frac{\xi_{\perp}^{\pm} - \rho_S \partial \psi_S^{\gamma} / \partial \gamma^{\pm}}{\tau_D^{\pm}} \right\rangle \operatorname{sgn}\left(\xi_{\perp}^{\pm} - \rho_S \partial \psi_S^{\gamma} / \partial \gamma^{\pm}\right) \\ &= \frac{\dot{\gamma}_0^{\pm}}{\tau_D^{\pm}} \left(\xi_{\perp}^{\pm} - \rho_S \frac{\partial \psi_S^{\gamma}}{\partial \gamma^{\pm}}\right). \end{aligned} \tag{4.32}$$

Rate-independent The dislocation motion is intrinsically rate-dependent. For numerical applications, however, it is useful to consider the special case of a rate-independent behavior. In this context, the Clausius-Duhem inequality in singular points according to Eq. (4.30) serves as motivation for the following yield conditions

$$f_S^{\pm} = \left(\xi_{\perp}^{\pm} - \rho_S \frac{\partial \psi_S^{\gamma}}{\partial \gamma^{\pm}}\right). \tag{4.33}$$

The corresponding Kuhn-Tucker conditions are given by

$$f_S^{\pm} \leq 0, \quad \dot{\gamma}^{\pm} \geq 0, \quad f_S^{\pm} \dot{\gamma}^{\pm} = 0, \tag{4.34}$$

cf. Bayerschen (2017, p. 93), Bayerschen et al. (2015, p. 8) and Wulfinghoff (2014, p. 116). Regarding single slip, no active-set search for the activated slip systems has to be considered. According to Eq. (4.34)$_3$, the GB is assumed to deform dissipation free, i.e., the entropy production on the GB vanishes. In this context, the GB is referred to as purely energetic and, thus, elastic. If the yield criterion is fulfilled, i.e., $f_S^{\pm} = 0$ hold true, the following boundary conditions for the plastic slip at the GB are obtained

$$\xi_{\perp}^{\pm} = \rho_S \frac{\partial \psi_S^{\gamma}}{\partial \gamma^{\pm}}. \tag{4.35}$$

Equation (4.35) explicitly prescribes the gradient of the plastic slip at the GB. The density of geometrically necessary dislocations is directly correlated to the gradient of the plastic slip, cf. Ashby (1970, Eq. (1.1)). Thus, an information on the dislocation pile-up at the GB is implicitly given. The obtained boundary conditions are structurally equivalent to those given by Özdemir and Yalcinkaya (2014, Eq. (40)). Moreover, the same boundary conditions can be obtained by requiring an isothermal behavior and a homogeneous temperature distribution, and by neglecting any effects due to heat supply, cf. Eq. (4.39).

Micro-free and micro-hard grain boundaries A micro-free GB does not resist against the movement of dislocations and, thus, the plastic slip. It is characterized by $\psi_S^{\gamma} = 0$ and $\xi_{\perp}^{\pm} = 0$, cf. Wulfinghoff and Böhlke (2012, p. 2696). In contrast, a micro-hard GB, cf. Aifantis et al. (2009); Bayerschen et al. (2015), completely prevents the flow of dislocations at the GB. Thus, the plastic slip at the GB vanishes and the corresponding boundary conditions are given by $\gamma^{\pm} = 0$.

Vanishing dissipation in singular points Similar to the derivations for regular points, $\dot{e}_S = \dot{\psi}_S + \dot{\theta}\eta_S + \theta\dot{\eta}_S$ is obtained. Due to the potential relation $\eta_S = -\partial\psi_S^\theta/\partial\theta$, cf. Eq. (4.29), the material time derivative of the Legendre transformation can be written as

$$\dot{e}_S = \dot{\psi}_S + \dot{\theta}\left(\eta_S + \theta\partial\eta_S/\partial\theta\right). \tag{4.36}$$

Consequently, $\dot{e}_S = \dot{\psi}_S$ holds true for an isothermal material behavior, i.e., $\dot{\theta} = 0$. Since a homogeneous temperature distribution is assumed, the jump of the temperature flux vanishes as well, i.e., $\{\boldsymbol{q}\} = 0$. Moreover, effects due to heat supply are neglected, i.e., $r_S = 0$. Thus, the local form of the balance of internal energy in singular points according to Eq. (4.23) reads

$$\rho_S\dot{\psi}_S - \{\boldsymbol{\xi}\dot{\gamma}\}\cdot\boldsymbol{n}_S = 0. \tag{4.37}$$

In addition, the dissipation inequality according to Eq. (4.25) reads

$$\rho_S\delta_S = -\rho_S\dot{\psi}_S + \{\boldsymbol{\xi}\dot{\gamma}\}\cdot\boldsymbol{n}_S \geq 0. \tag{4.38}$$

Comparing Eq. (4.37) with Eq. (4.38) illustrates that the dissipation on the singular surface vanishes under the met assumptions, which corresponds to the concept of an ideal wall, cf. Müller (1985). Accounting for Eq. (4.26) as well as for Eq. (4.27) and neglecting the thermal contriubtion, Eq. (4.37) can be written as

$$\left(\xi_\perp^+ - \rho_S\frac{\partial\psi_S^\gamma}{\partial\gamma^+}\right)\dot{\gamma}^+ + \left(\xi_\perp^- - \rho_S\frac{\partial\psi_S^\gamma}{\partial\gamma^-}\right)\dot{\gamma}^- = 0. \tag{4.39}$$

In order to fullfill Eq. (4.39), the terms in each bracket could be assumed to vanish leading to the boundary conditions for the plastic slip at the GB given by Eq. (4.35). Finally, a purely energetic GB is obtained as a consequence of the above mentioned restrictions.

4.2 Connection to an Extended Principle of Virtual Power

4.2.1 Weak forms

Balance of linear momentum To compare the presented framework with an extended principle of virtual power, the weak forms of the balance of linear momentum, cf. Eq. (3.12), and the PDE often referred to as micro-force balance, cf. Eq. $(3.33)_1$, are used. To this end, the balance of linear momentum is multiplied by a test funtion $\boldsymbol{f}$, which is conmtinuous across $\mathcal{S}$, and integrated over the volume $\mathcal{V}$. Moreover, the divergence theorem according to Eq. (4.4) is applied, leading to

$$\begin{aligned}\int\limits_{\mathcal{V}^+\cup\mathcal{V}^-} &\rho\,(\boldsymbol{a}-\boldsymbol{b})\cdot\boldsymbol{f}+\boldsymbol{\sigma}\cdot\mathrm{grad}\,(\boldsymbol{f})\;\mathrm{d}v\\ &-\int\limits_{\mathcal{F}^+\cup\mathcal{F}^-} \boldsymbol{t}\cdot\boldsymbol{f}\,\mathrm{d}a+\int\limits_{\mathcal{S}} [\![\boldsymbol{t}\cdot\boldsymbol{f}]\!]\,\mathrm{d}a=0.\end{aligned} \tag{4.40}$$

Subsequently, a quasi-static process without body forces is considered, i.e.,

$$\boldsymbol{a}=\boldsymbol{0},\quad \boldsymbol{b}=\boldsymbol{0},\quad \boldsymbol{b}_{\mathcal{S}}=\boldsymbol{0}. \tag{4.41}$$

Moreover, recalling the balance of linear momentum in singular points according to Eq. (4.22), Eq. (4.40) yields

$$\int\limits_{\mathcal{V}^+\cup\mathcal{V}^-} \boldsymbol{\sigma}\cdot\mathrm{grad}\,(\boldsymbol{f})\;\mathrm{d}v-\int\limits_{\mathcal{F}^+\cup\mathcal{F}^-} \boldsymbol{t}\cdot\boldsymbol{f}\,\mathrm{d}a=0. \tag{4.42}$$

Micro-force balance Multiplication of Eq. $(3.33)_1$ by a test function f, integration over $\mathcal{V}$ and application of the divergence theorem according to Eq. (4.4) yields

$$
-\int\limits_{\mathcal{V}^+\cup\mathcal{V}^-} \pi f + \boldsymbol{\xi}\cdot\operatorname{grad}(f)\,\mathrm{d}v
+\int\limits_{\mathcal{F}^+\cup\mathcal{F}^-} (\boldsymbol{\xi}\cdot\boldsymbol{n})f\,\mathrm{d}a - \int\limits_{\mathcal{S}} [\![\boldsymbol{\xi} f]\!]\cdot\boldsymbol{n}_{\mathcal{S}}\,\mathrm{d}a = 0. \tag{4.43}
$$

Considering Eq. (3.22) and Eq. (4.26), Eq. (4.43) can be written as

$$
-\int\limits_{\mathcal{V}^+\cup\mathcal{V}^-} \pi f + \boldsymbol{\xi}\cdot\operatorname{grad}(f)\,\mathrm{d}v
+\int\limits_{\mathcal{F}^+\cup\mathcal{F}^-} \tilde{t} f\,\mathrm{d}a - \int\limits_{\mathcal{S}} \xi_\perp^+ f^+ + \xi_\perp^- f^-\,\mathrm{d}a = 0. \tag{4.44}
$$

4.2.2 Extended principle of virtual power

Virtual rates The weak forms according to Eq. (4.42) and Eq. (4.44) are used to formulate an extended principle of virtual power. To this end, the test functions $\boldsymbol{f}$ and f are replaced by the corresponding virtual rates $\delta\boldsymbol{v}$ and $\delta\dot{\gamma}$, respectively.

Comparison to models used in the literature Using the virtual rates $\delta\boldsymbol{v}$ and $\delta\dot{\gamma}$, the sum of Eq. (4.42) and Eq. (4.44) yields the principle of virtual power as $\delta\mathcal{P}_{\mathrm{int}} = \delta\mathcal{P}_{\mathrm{ext}}$ with

$$
\begin{aligned}
\delta\mathcal{P}_{\mathrm{int}} &= \int\limits_{\mathcal{V}^+\cup\mathcal{V}^-} \pi\delta\dot{\gamma} + \boldsymbol{\xi}\cdot\operatorname{grad}(\delta\dot{\gamma}) + \boldsymbol{\sigma}\cdot\operatorname{grad}(\delta\boldsymbol{v})\,\mathrm{d}v \\
&\quad + \int\limits_{\mathcal{S}} \xi_\perp^+\delta\dot{\gamma}^+ + \xi_\perp^-\delta\dot{\gamma}^-\,\mathrm{d}a, \\
\delta\mathcal{P}_{\mathrm{ext}} &= \int\limits_{\mathcal{F}^+\cup\mathcal{F}^-} \tilde{t}\,\delta\dot{\gamma} + \boldsymbol{t}\cdot\delta\boldsymbol{v}\,\mathrm{d}a.
\end{aligned} \tag{4.45}
$$

This extended principle of virtual power equals structurally the formulation given by Erdle and Böhlke (2017, Eq. (10)) regarding single slip.

4.3 Interim Conclusions

Regarding small deformations and single slip, thermodynamically consistent flow rules of the plastic slip on the GB are provided. To this end, invariance considerations of an extended energy balance with respect to a change of observer are exploited. The framework discussed is based on the assumptions listed below.

- The GB is considered as a material singular surface.
- The GB separates two grains from each other.
- The grains are considered as material volumes.
- The temperature is continuous across the GB.
- The energy balance is supplemented by an additional flux term across the boundary of the material volumes.
- The additional flux term is associated with an additional scalar-valued DOF, representing the plastic slip.
- No energy flux is present across the boundary of the GB.
- Contrarily to, e.g., Eringen (1999), the energy balance considered here accounts for contributions defined on the singular surface such as a specific internal energy and a specific kinetic energy.
- Effects due to micro-inertia and generalized micro-body forces are neglected.
- The balance of energy is invariant with respect to a change of observer.

The results of the present chapter are summarized below.

- Invariance considerations regarding the extended energy balance yield the conservation of mass as well as the balance of linear momentum and internal energy in singular points.

- In singular points, the mean curvature of the GB is taken into account by the balance of mass, explicitly. Contrarily, the balance of linear momentum and internal energy depend only implicitly on the curvature of the GB by means of the balance of mass.
- If an isothermal material behavior with a homogeneous temperature distribution is considered, and effects due to heat supply are neglected, the dissipation vanishes at the GB. Thus, the behavior of the GB is purely elastic.
- Taking into account the obtained balance equations, the exploitation of the Coleman-Noll procedure yields the flow rule for the plastic slip at the GB.
- A classification of thermodynamically consistent flow rules for the plastic slip at the GB is provided.
- The equivalence between an extended energy balance and an extended principle of virtual power is shown.

Chapter 5

Application to a Slip Gradient Crystal Plasticity[1]

Motivation The framework, presented in the previous two chapters, is now applied to a slip gradient crystal plasticity. In this context, the distribution of the plastic slip with respect to a two- and a three-phase laminate material is presented by means of analytical solutions. The elastoplastic phase mimic a grain of an FCC crystal with one active slip system. The adjacent elastoplastic phases of the three-phase laminate represent the grains of a bicrystal. The effect of the internal length scale on the distribution of the plastic slip is considered. A grain boundary effect is discussed which is based on a variation of the internal length scale in one of the two elastoplastic phases.

5.1 General Remarks on Crystal Plasticity

5.1.1 Deformation behavior

Small deformations Subsequently, small deformations are considered, for brevity. Thus, the stress power $\boldsymbol{\sigma} \cdot \boldsymbol{D}$ is given by $\boldsymbol{\sigma} \cdot \dot{\varepsilon}$. Here, ε denotes

[1] Most of the content of this chapter is taken directly from the articles Prahs and Böhlke (2019b;a). Additional content concerning the influence of the defect parameter, the inital hardening modulus, the grain width and the orientation of the slip system on the distribution of the plastic slip is considered. Minor linguistic changes and abbreviations have been made.

the infinitesimal strain tensor, defined as the symmetric gradient of the displacement $\boldsymbol{u}$, i.e., $\boldsymbol{\varepsilon} = \mathrm{sym}(\mathrm{grad}\,(\boldsymbol{u}))$. An additive decomposition of the strain $\boldsymbol{\varepsilon}$ into an elastic $\boldsymbol{\varepsilon}^{\mathrm{e}}$ and a plastic contribution $\boldsymbol{\varepsilon}^{\mathrm{p}}$ is assumed.

5.1.2 Crystal plasticity

Plastic slip as additional degree of freedom In the context of a slip gradient crystal plasticity theory, the plastic slip γ within a slip system is considered as additional DOF. Thus, the additional energy flux $\tilde{t}\dot{\gamma}$, cf. Eq. (4.3), is associated with the movement of dislocations. Single slip is considered for simplicity in the following. Consequently, the rate $\dot{\gamma}$ is regarded as plastic slip rate while the vector field $\boldsymbol{\xi}$ is considered as corresponding gradient stress vector. Additionally, it is assumed that the plastic part $\boldsymbol{\varepsilon}^{\mathrm{p}}$ of the strain tensor and its rate are given by $\boldsymbol{\varepsilon}^{\mathrm{p}} = \gamma\,\boldsymbol{M}$ and $\dot{\boldsymbol{\varepsilon}}^{\mathrm{p}} = \dot{\gamma}\,\boldsymbol{M}$, respectively, with the corresponding Schmid tensor $\boldsymbol{M} = \mathrm{sym}(\boldsymbol{d} \otimes \boldsymbol{n})$. Here, the slip direction is denoted as $\boldsymbol{d}$ and the slip plane normal as $\boldsymbol{n}$. The scalar product between the Cauchy stress and the Schmid tensor yields the resolved shear stress $\tau = \boldsymbol{\sigma} \cdot \boldsymbol{M}$, cf., e.g., Bertram (2005).

Grain boundary as material singular surface In the context of a slip gradient crystal plasticity theory, the material volumes $\mathcal{V}^{+}$ and $\mathcal{V}^{-}$, as depicted in Fig. 4.1, can be considered as grains of a grain aggregate, and the material singular surface $\mathcal{S}$ as a grain boundary (GB) separating two grains from each other. Consequently, quantities indicated by a $'+'$-sign are associated with $\mathcal{V}^{+}$, quantities indicated by a $'-'$-sign to $\mathcal{V}^{-}$.

5.2 Bulk Material - Regular Points

5.2.1 Application to a two-phase laminate material

Two-phase laminate material An analytical, one-dimensional solution of Eq. (3.34) is discussed in the context of a laminate material consisting of

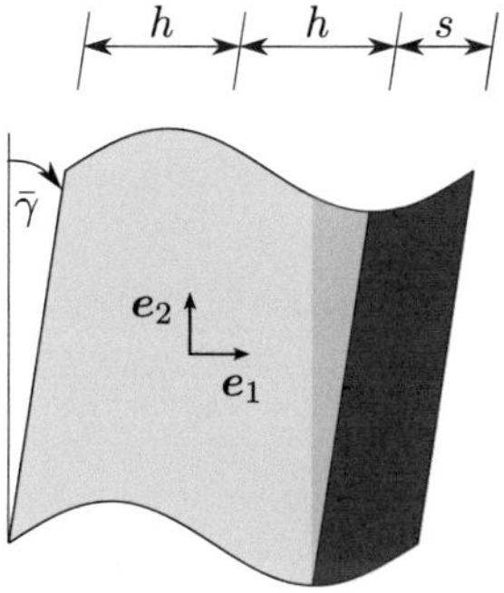

Figure 5.1: Considered laminate material subjected to shear in e_1- and periodic fluctuation in e_2-direction. The elastoplastic phase is illustrated in light gray, the elastic phase in dark gray. The coordinate system is located in the center of the elastoplastic phase.

two phases. The material behavior of one phase is assumed to be purely elastic. Contrarily, the material behavior of the second phase is considered to be elastoplastic. The plastic behavior is characterized by an individual slip system of a face-centered cubic (FCC) single crystal. A schematic illustration of the considered laminate is given in Fig. 5.1. The elastic phase is illustrated in dark gray, the elastoplastic phase in light gray. The normal vector and the slip direction of the considered slip system are given by $\boldsymbol{n} = \boldsymbol{e}_2$ and $\boldsymbol{d} = \boldsymbol{e}_1$. The coordinate system is located within the center of the elastoplastic phase. While the elastoplastic phase has a width of $2h$, the width of the elastic phase is s.

Specific free energy and material parameters The plastic slip γ cannot be measured and, thus, is not an internal state variable, cf. Maugin (1992, p. 277). Regarding monotonic loading, the evolution of the plastic slip is closely related to the evolution of the dislocation density by the Kocks-Mecking law, cf. Kocks (1976). For single slip, this law establishes a relation between the plastic slip within the slip system and the corresponding dislocation density, if the slip rate is considered constant, cf. Bayerschen (2017, p. 53). Thus, the plastic slip can nevertheless be used as internal state variable in an approximate sense.

The contributions to the specific free energy are given by

$$\begin{aligned} \psi_{\mathrm{e}}(\varepsilon - \varepsilon^{\mathrm{p}}) &= \frac{1}{2\rho}\left(\varepsilon - \varepsilon^{\mathrm{p}}\right) \cdot \left(\mathbb{C}\left[\varepsilon - \varepsilon^{\mathrm{p}}\right]\right), \\ \psi_{\mathrm{g}}(\nabla_{\mathrm{n}}\gamma) &= \frac{1}{2\rho} K_g \nabla_{\mathrm{n}}\gamma \cdot \nabla_{\mathrm{n}}\gamma, \\ \psi_{\mathrm{h}}(\gamma) &= \frac{1}{2\rho}\Theta_0 \gamma^2. \end{aligned} \tag{5.1}$$

Here, the elastic stiffness tensor of fourth order is referred to as $\mathbb{C}$ and the initial hardening modulus as Θ_0. The defect parameter K_g is inversely proportional to the stored initial dislocation density, referred to as ρ_0, cf. Bayerschen et al. (2016b), and, thus, introduces an internal length scale to the model, cf. Wulfinghoff et al. (2013). The planar gradient with respect to the considered slip system is denoted as ∇_{n}. Accounting for the lattice stretch and rotation, the elastic contribution ψ_{e} is assumed to be quadratic in the difference of the total strain ε and the plastic strain ε^{p}. While the specific defect energy ψ_{g} is related to hardening on the basis of geometrical necessary dislocations, the hardening contribution ψ_{h} accounts for hardening due to statistically stored dislocations in the context of monotonic loadings, cf. Bayerschen et al. (2015). Subsequently, an analytical solution of Eq. (3.34) is discussed. To obtain a linear differential equation, the defect contribution ψ_{g} as well as the hardening contribution ψ_{h} are considered to be quadratic. Regarding a numerical implementation of the presented theory, more complex contributions are feasible such as Voce-hardening, cf. Wulfinghoff et al. (2013), latent hardening, cf. Ortiz and Repetto (1999), or a power-law defect energy, cf. Bardella (2010); Bayerschen and Böhlke (2016). Assuming a power-law for the defect contribution ψ_{g} yields gradients at the GB that are more in line with results obtained by discrete dislocation dynamics simulations, cf. Bayerschen et al. (2015). Thermal effects are neglected, i.e.,the thermal contribution $\psi_\theta(\theta)$ vanishes. The density of mass is referred to as ρ. In the following, it is assumed that the stiffness tensor, the defect parameter and the initial hardening modulus are constant parameters. Assuming the elastic material behavior to be isotropic, the elastic constants are chosen to be $G = 27$ GPa and $\nu = 0.347$, representing the elastic behavior of aluminum. Moreover, the defect parameter and the initial hardening modulus are chosen as $K_g = 84$ µN and $\Theta_0 = 1075$ MPa, respectively. The material parameters are in line with Bayerschen et al. (2015). Under consideration of the contributions to the specific free energy, cf. Eq. (5.1), Eq. (3.34) reads

$$\Delta_{\mathrm{n}}\gamma - \gamma\frac{\Theta_0}{K_g} = -\frac{\tau}{K_g}, \tag{5.2}$$

were Δ_{n} denotes the Laplacian that uses the planar gradient.

5.2.2 Solution of the considered boundary value problem

Kinematics Subsequently, the following ansatz for the displacement field $\boldsymbol{u}(\boldsymbol{x})$ is considered

$$\boldsymbol{u} = \bar{\gamma}x_2\boldsymbol{e}_1 + \tilde{u}(x_1)\boldsymbol{e}_2. \tag{5.3}$$

This ansatz is an approximation if a small elastic phase is considered; it is related to Forest (2013); Forest and Guéninchault (2013), Wulfinghoff et al. (2015). Here, $\bar{\gamma}$ denotes the constant macroscopic shear and $\tilde{u}(x_1)$ a periodic fluctuation. The corresponding infinitesimal strain tensor and the plastic strain tensor are given by

$$\boldsymbol{\varepsilon} = \frac{1}{2}\left(\bar{\gamma} + \tilde{u}'(x_1)\right)\left(\boldsymbol{e}_1 \otimes \boldsymbol{e}_2 + \boldsymbol{e}_2 \otimes \boldsymbol{e}_1\right), \tag{5.4}$$

$$\boldsymbol{\varepsilon}^{\mathrm{p}} = \frac{1}{2}\gamma(x_1)\left(\boldsymbol{e}_1 \otimes \boldsymbol{e}_2 + \boldsymbol{e}_2 \otimes \boldsymbol{e}_1\right). \tag{5.5}$$

Based on the applied deformation, the plastic slip depends purely on x_1. The derivative of a quantity with respect to x_1 is denoted by $(\cdot)'$. The elastic strain $\boldsymbol{\varepsilon}^{\mathrm{e}} = \boldsymbol{\varepsilon} - \boldsymbol{\varepsilon}^{\mathrm{p}}$ is obtained by the assumed additive decomposition of the infinitesimal strain tensor in an elastic and a plastic contribution. For brevity, an isotropic elastic behavior is assumed in the following. Hooke's law for linear elasticity yields the corresponding Cauchy stress $\boldsymbol{\sigma}$ as

$$\boldsymbol{\sigma} = G\left(\bar{\gamma} + \tilde{u}'(x_1) - \gamma(x_1)\right)\left(\boldsymbol{e}_1 \otimes \boldsymbol{e}_2 + \boldsymbol{e}_2 \otimes \boldsymbol{e}_1\right), \tag{5.6}$$

with the shear modulus G. The balance of linear momentum, given by Eq. (3.12), yields the differential equation

$$\tilde{u}''(x_1) = \gamma'(x_1). \tag{5.7}$$

Regarding Eq. (5.2), the Laplacian $\Delta_{\mathrm{n}}\gamma$ can be replaced by $\gamma''(x_1)$. Moreover, the resolved shear stress $\tau = G\left(\bar{\gamma} + \tilde{u}'(x_1) - \gamma(x_1)\right)$ can be reformulated by means of Eq. (5.7). Thus, Eq. (5.2) reads

$$\gamma'' - \frac{\Theta_0}{K_g}\gamma = -\frac{\sigma_0}{K_g}, \qquad \sigma_0 = G\left(\bar{\gamma} + c\right), \tag{5.8}$$

where c denotes the integration constant if Eq. (5.7) is integrated once with respect to x_1. At the boundaries between both phases, the plastic slip vanishes, i.e.,$\gamma\left(-h\right) = 0$ and $\gamma\left(h\right) = 0$ hold true. The function $\tilde{u}(x_1)$ is considered to be a periodic fluctuation. Thus,

$$\int_{-h}^{h+s} \tilde{u}(x_1)\,\mathrm{d}x_1 = 0, \qquad \int_{-h}^{h+s} \tilde{u}'(x_1)\,\mathrm{d}x_1 = 0 \tag{5.9}$$

hold true, cf. Wulfinghoff et al. (2015, eq. (27)). These conditions are used to determine the integration constants that arise in the context of Eq. (5.7). Finally, solving Eq. (5.8) closes the ansatz for the displacement field, given by Eq. (5.3).

Solution The ordinary, linear differential equation, cf. Eq. (5.8), can be solved analytically. The solution is given in dependency of σ_0 and reads

$$\gamma(x_1) = -\frac{\sigma_0}{\Theta_0}\left(\mathrm{e}^{\sqrt{\frac{\Theta_0}{K_g}}x_1} + \mathrm{e}^{-\sqrt{\frac{\Theta_0}{K_g}}x_1}\right)\left(\mathrm{e}^{\sqrt{\frac{\Theta_0}{K_g}}h} + \mathrm{e}^{-\sqrt{\frac{\Theta_0}{K_g}}h}\right)^{-1} + \frac{\sigma_0}{\Theta_0}. \tag{5.10}$$

For brevity, the integration constants to determine $\tilde{u}(x_1)$ are not given explicitly, here. They are documented in the appendix, cf. Eq. (A.11) and Eq. (A.12). Regarding a sequence of equilibrium states, the distribution of the plastic slip $\gamma(x_1)$ is depicted in Fig. 5.2. The amplitude σ_0 is chosen as 15 MPa, 25 MPa and 35 MPa, respectively. For the illustration, h is chosen as $0.5\mu\mathrm{m}$. The solution according to Eq. (5.10) is shown in purple. Due to the quadratic defect energy, a parabolic distribution of the plastic slip is obtained. The analytical solution is qualitatively in line with the numerical results presented in Bayerschen et al. (2015). Neglecting the contribution due to the defect energy leads to a constant distribution of the plastic slip. This solution represents the classical distribution of the plastic slip without gradient effects and is illustrated in Fig. 5.2 in orange. The absolute value of the classical distribution of the plastic slip is significantly higher compared to the parabolic distribution.

5.2.3 Influence of the defect parameter and the initial hardening modulus on the solution

Effect of the defect parameter Subsequently, the influence of the defect parameter K_g on the distribution of the plastic slip is illustrated for $\sigma_0 = 35$ MPa. It is depicted in Fig. 5.3. To this end, the defect parameter is varied within an interval of $[0\ \mu\mathrm{N}, 840\ \mu\mathrm{N}]$. The choice of $K_g = 0\ \mu\mathrm{N}$ corresponds to a vanishing defect energy. Consequently, the distribution of the plastic slip is constant throughout the plastic phase. This solution,

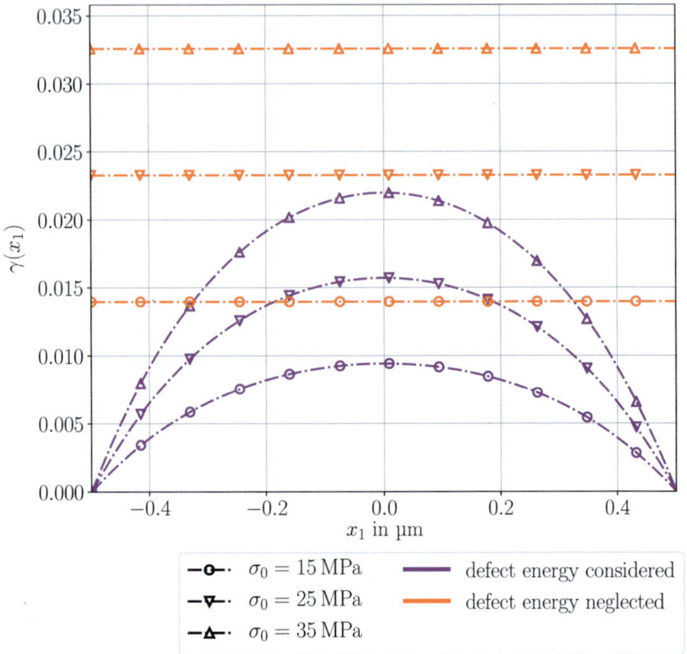

Figure 5.2: Parabolic distribution of the plastic slip $\gamma(x_1)$ as a consequence of the quadratic defect energy. Three different amplitudes σ_0 are considered. If the defect energy is neglected, the solution for the plastic slip is homogeneous.

referred to as classical distribution, is given by the second term of Eq. (5.10). The first term is based on the non-vanishing defect energy. Thus, the classical contribution constitutes an upper bound for the distribution of the plastic slip, here. Regarding the limiting case $K_g \to \infty$, the plastic slip vanishes, i.e., $\lim_{K_g \to \infty} \gamma(x_1) = 0$. Consequently, the plastic slip takes values in the interval $[0, \Theta_0/\sigma_0]$. The absolute value of the plastic slip decreases for an increase of the defect parameter. While the distribution of the plastic slip has a distinct plateau for small values of K_g, the distribution becomes parabolic for higher values of the defect parameter. The higher the value of the defect parameter, the smaller the absolut value of γ' becomes at the transition to the purely elastic phases.

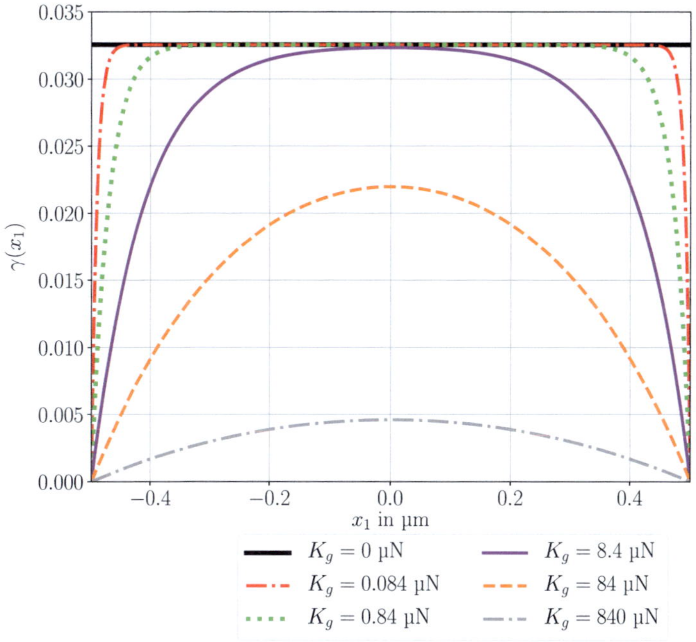

Figure 5.3: The classical distribution of the plastic slip constitutes an upper bound. Regarding an increasing defect parameter, the plastic slip decreases compared to the classical distribution. Increasing the defect parameter yields a decrease of the absolute value of gradient of the plastic slip at the GB. The plastic slip shows a plateau-like behavior for small values of the defect parameter. For increasing defect parameter, the distribution of the plastic slip is of parabolic shape.

Effect of the initial hardening modulus The influence of the initial hardening modulus Θ_0 on the distribution of the plastic slip is illustrated for $\sigma_0 = 35$ MPa and $K_g = 84$ µN in Fig. 5.4. Here, Θ_0 is varied within the interval $[0 \text{ MPa}, 10750 \text{ MPa}]$. If the initial hardening modulus is neglected, the ordinary differential equation (ODE) according to Eq. (5.8) is simplified and the distribution of the plastic slip is given by

$$\gamma(x_1) = \frac{\sigma_0}{K_g} \left(h^2 - x_1^2\right). \tag{5.11}$$

This distribution constitutes an upper bound for the plastic slip. Increasing values of Θ_0 lead to a decrease of the absolute value of the plastic slip. At the transition to the elastic phase, the absolute value of γ' is highest for $\Theta_0 = 0$. An increase of Θ_0 yields a decrease of the absolute value of the gradient of the plastic slip. The choice of the initial hardening modulus does not affect the parabolic shape of the distribution of the plastic slip.

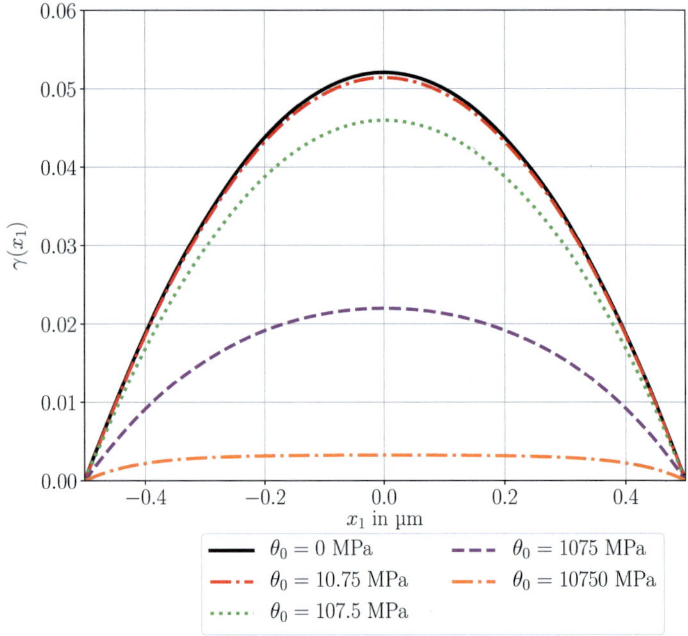

Figure 5.4: Regarding different values of the initial hardening modulus, an upper bound is given by $\theta_0 = 0$ MPa. An increasing value of θ_0 yields a decrease of the value of the plastic slip. Moreover, an increase of the initial hardening modulus yields a decrease of the absolute value of the gradient of the plastic slip at the GB. The distribution of the plastic slip exhibits a parabolic profile for all considered choices of θ_0.

5.2.4 Orientation of the slip system

Kinematics In contrast to the previous example, a rotated slip system with the slip plane normal $\boldsymbol{n} = n_1 \boldsymbol{e}_1 + n_2 \boldsymbol{e}_2$ and the slip direction $\boldsymbol{d} = d_1 \boldsymbol{e}_1 + d_2 \boldsymbol{e}_2$ is now considered. Both, $\boldsymbol{n}$ and $\boldsymbol{d}$ are unit vectors. The plastic strain tensor reads

$$\begin{aligned} \boldsymbol{\varepsilon}^{\mathrm{p}} =& \gamma \left(n_1 d_1 \, \boldsymbol{e}_1 \otimes \boldsymbol{e}_1 + n_2 d_2 \, \boldsymbol{e}_2 \otimes \boldsymbol{e}_2 \right) \\ &+ \frac{\gamma}{2} \left(n_1 d_2 + n_2 d_1 \right) \left(\boldsymbol{e}_1 \otimes \boldsymbol{e}_2 + \boldsymbol{e}_2 \otimes \boldsymbol{e}_1 \right) . \end{aligned} \tag{5.12}$$

The infinitesimal strain tensor $\boldsymbol{\varepsilon}$ is given unalteredly by Eq. (5.4).

Exploitation of the balance of linear momentum Taking into accout Eq. (5.12), the balance of linear momentum according to Eq. (3.12) yields the differential equations

$$0 = -\frac{\partial \gamma}{\partial x_1} n_1 d_1 - \frac{1}{2} \frac{\partial \gamma}{\partial x_2} \left(n_1 d_2 + n_2 d_1 \right), \tag{5.13}$$

$$0 = \frac{1}{2} \frac{\mathrm{d}^2 \tilde{u}}{\mathrm{d} x_1^2} - \frac{1}{2} \frac{\partial \gamma}{\partial x_1} \left(n_1 d_2 + n_2 d_1 \right) - \frac{\partial \gamma}{\partial x_2} n_2 d_2 . \tag{5.14}$$

Consequently, the balance of linear momentum does not yield a restriction of the dependency of the plastic slip on x_1 as in the previous example. Thus, the PDE according to Eq. (5.2) cannot be simplified to an ODE similar to Eq. (5.8). Solving the PDE by analytical means to obtain the distribution of the plastic slip can get quite involved. To this end, a numerical solution of the PDE is suggested.

5.3 Grain Boundary - Singular Points

5.3.1 Application to a three-phase laminate material

Three-phase laminate material The considered three-phase laminate material is schematically illustrated in Fig. 5.5. Two of the three phases are considered to behave elastoplastic representing the grains of a bicrystal. The left grain is referred to as $\mathcal{V}^-$ and the right grain as $\mathcal{V}^+$. An individual slip system of an FCC crystal characterizes the plastic behavior within each grain. The normal and slip directions of the considered coherent slip systems are given by $\boldsymbol{n} = \boldsymbol{e}_2$ and $\boldsymbol{d} = \boldsymbol{e}_1$. The third phase is purely elastic. Thus, micro-hard boundary conditions regarding the plastic slip are present at the transition between the elastic and an elastoplastic phase. A GB with the normal vector $\boldsymbol{n}_S = \boldsymbol{e}_1$ separates both elastoplastic phases. While the width of the left and the right grain is referred to as h^- and h^+, respectively, the width of the elastic phase is denoted as s. The origin of the coordinate system lies on the GB.

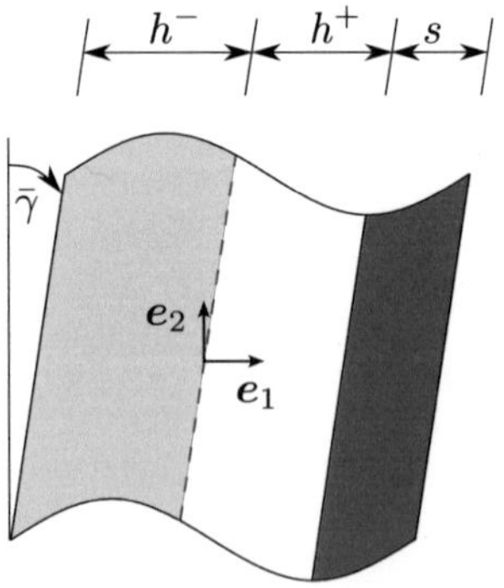

Figure 5.5: Considered three-phase laminate material subjected to shear in $\boldsymbol{e}_1$- and periodic fluctuation in $\boldsymbol{e}_2$-direction. The elastoplastic phases are represented by the light gray and the white areas. The elastic phase is illustrated in dark gray. Both elastoplastic phases are separated by a GB, depicted as dashed, red line. The coordinate system is located on the GB.

Specific free energy of the bulk material Regarding the bulk material, the contributions to the specific free energy are given by Eq. (5.1). Moreover, an isothermal behavior is considered, i.e., the thermal contribution ψ_θ is neglected. While the elastic and the hardening contribution are chosen to be the same throughout $\mathcal{V}^+$ and $\mathcal{V}^-$, the defect parameter is assumed to differ in both grains. Consequently, the defect parameter with respect to $\mathcal{V}^+$ is denoted by K_g^+ and with respect to $\mathcal{V}^-$ as K_g^-. Throughout the work at hand, the quantities $\mathbb{C}$, Θ_0 and $K_g^\pm$ are considered constant. For brevity, the elastic material behavior is considered to be isotropic with the shear modulus $G = 27$ GPa and Poisson's ratio $\nu = 0.35$, representing the elastic behavior of aluminum, cf. Bayerschen et al. (2015). The initial hardening modulus is chosen as $\Theta_0 = 1075\,\mathrm{MPa}$. The material parameters G, ν and Θ_0 are chosen according to Bayerschen et al. (2015). Accounting for the contributions according to Eq. (5.1), the PDE given by Eq. (3.34) reads

$$\Delta_{\mathrm{n}}\gamma^\pm - \frac{\Theta_0}{K_g^\pm}\gamma^\pm = -\frac{\tau^\pm}{K_g^\pm}. \tag{5.15}$$

The resolved shear stress in the respective grain is referred to as τ^- and τ^+. In case of coherent slip systems, the normal and slip directions of the slip systems are identical in both grains, and briefly referred to as $\boldsymbol{n}$ and $\boldsymbol{d}$ instead of $\boldsymbol{n}^\pm$ and $\boldsymbol{d}^\pm$. Thus, $\tau^\pm = \boldsymbol{\sigma}^\pm \cdot \boldsymbol{M}$ holds true.

Specific free energy of the grain boundary Subsequently, the special case of energetic grain boundaries, characterized by Eq. (4.35), is considered. Based on the GB burgers tensor $\boldsymbol{G}$, cf. Gurtin (2008), the GB contribution ψ_S^γ can be formulated as a quadratic form reading

$$\psi_S^\gamma = \frac{\kappa}{2\,\rho_S}\|\boldsymbol{G}\|^2. \tag{5.16}$$

Here, κ denotes the GB strength, cf., e.g., Özdemir and Yalcinkaya (2014). It is chosen as $\kappa = 100\,\mathrm{N/m}$ similar to Wulfinghoff et al. (2013). The expression $\|G\|$ can be formulated in terms of the so-called interaction moduli. Regarding single slip and the more general case of non-coherent slip systems, the simplified intra-grain interaction moduli $C^{\pm\pm}$ and the inter-grain interaction moduli C^{-+} are given by

$$\begin{aligned} C^{\pm\pm} &= \|\boldsymbol{n}^{\pm} \times \boldsymbol{n}_{\mathcal{S}}\|^2, \\ C^{-+} &= \left(\boldsymbol{d}^{-} \cdot \boldsymbol{d}^{+}\right) \left(\boldsymbol{n}^{-} \times \boldsymbol{n}_{\mathcal{S}}\right) \cdot \left(\boldsymbol{n}^{+} \times \boldsymbol{n}_{\mathcal{S}}\right), \end{aligned} \tag{5.17}$$

cf. Gurtin (2008); Özdemir and Yalcinkaya (2014). Thus, the GB contribution $\psi_{\mathcal{S}}^{\gamma}$ can be written in terms of the interaction moduli, reading

$$\psi_{\mathcal{S}}^{\gamma}\left(\gamma^{+},\gamma^{-}\right) = \frac{\kappa}{2\,\rho_{\mathcal{S}}}\left(C^{++}\gamma^{+2} + C^{--}\gamma^{-2} - 2C^{-+}\gamma^{-}\gamma^{+}\right). \tag{5.18}$$

Here, the slip system considered in $\mathcal{V}^{+}$ is specified by $\boldsymbol{n}^{+}$ and $\boldsymbol{d}^{+}$, the slip system in $\mathcal{V}^{-}$ by $\boldsymbol{n}^{-}$ and $\boldsymbol{d}^{-}$, respectively. While the intra-grain interaction moduli $C^{\pm\pm}$ depend on the slip plane normal of an individual grain, the inter-grain interaction moduli C^{-+} depend on the slip directions and slip plane normals of two adjacent grains. Moreover, both moduli depend on the normal of the GB. Thus, the intra-grain interaction moduli carry information about the orientation of slip systems with respect to the GB within an individual grain. Conversely, the inter-grain interaction moduli constitute a measure for the geometrical mismatch between slip systems of adjacent grains, cf. Özdemir and Yalcinkaya (2014). Moreover, the range of values is between 0 and 1, concerning the intra-grain interaction moduli $C^{\pm\pm}$, and between -1 and 1, regarding the interaction moduli C^{-+}. With respect to the range of values, three special cases are to be discussed. If $C^{-+} = C^{++} = C^{--}$ holds, then slip system α in grain $\mathcal{V}^{+}$ and slip system β in grain $\mathcal{V}^{-}$ are coherent. If $C^{++} = 0$ or $C^{--} = 0$ holds, the slip plane in $\mathcal{V}^{+}$ or $\mathcal{V}^{-}$ is parallel to the GB. Consequently, if at least one slip system is parallel to the GB, $C^{-+} = 0$ holds. From Eq. (5.17) it follows that the slip system in grain $\mathcal{V}^{-}$ or $\mathcal{V}^{+}$ is perpendicular to the GB if $C^{--} = 1$ or $C^{++} = 1$ holds, respectively.

Subsequently, the special case of coherent slip systems is discussed. In this context, C^{-+}, C^{++} and C^{--} are briefly replaced by C. As stated previously for the bulk material, thermal effects are also neglected with respect to the GB. Therefore, the thermal contribution $\psi_{\mathcal{S}}^{\theta}$ vanishes. Using the constitutive equation for the generalized stress $\boldsymbol{\xi}$, cf. Eq. (3.30), the left hand side of the boundary conditions for the plastic slip at the GB, cf. Eq. (4.35), yields

$$\xi_{\perp}^{\pm} = \boldsymbol{\xi}^{\pm}\big|_{\mathcal{S}} \cdot \boldsymbol{n}_{\mathcal{V}}^{\pm} \equiv \rho \left.\frac{\partial \psi_{\mathrm{g}}^{\pm}}{\partial \nabla_{\mathrm{n}} \gamma^{\pm}}\right|_{\mathcal{S}} \cdot \boldsymbol{n}_{\mathcal{V}}^{\pm}. \tag{5.19}$$

Regarding Eq. (5.19) and the quadratic specific defect energy according to Eq. (5.1), the boundary conditions following Eq. (4.35) read

$$\begin{aligned} \nabla_{\mathrm{n}}\gamma^{+}\big|_{\mathcal{S}} \cdot \boldsymbol{n}_{\mathcal{V}}^{+} &= \frac{\rho_{\mathcal{S}}}{K_g^{+}} \left.\frac{\partial \psi_{\mathcal{S}}^{\gamma}}{\partial \gamma^{+}}\right|_{\mathcal{S}} = \frac{C\kappa}{K_g^{+}} \left.(\gamma^{+} - \gamma^{-})\right|_{\mathcal{S}}, \\ \nabla_{\mathrm{n}}\gamma^{-}\big|_{\mathcal{S}} \cdot \boldsymbol{n}_{\mathcal{V}}^{-} &= \frac{\rho_{\mathcal{S}}}{K_g^{-}} \left.\frac{\partial \psi_{\mathcal{S}}^{\gamma}}{\partial \gamma^{-}}\right|_{\mathcal{S}} = \frac{C\kappa}{K_g^{-}} \left.(\gamma^{-} - \gamma^{+})\right|_{\mathcal{S}}. \end{aligned} \tag{5.20}$$

5.3.2 Solution of the considered boundary value problem

Formulation of the boundary value problem Regarding the displacement field $\boldsymbol{u}(\boldsymbol{x})$, the following ansatz is considered, subsequently

$$\begin{aligned} \boldsymbol{u}(x_1, x_2) &= \bar{\gamma} x_2 \boldsymbol{e}_1 + \tilde{u}(x_1)\boldsymbol{e}_2, \\ \tilde{u}(x_1) &= \begin{cases} \tilde{u}_{-}(x_1), & x \in [-h^{-}, 0) \\ \tilde{u}_{+}(x_1), & x \in [0, h^{+} + s] \end{cases}. \end{aligned} \tag{5.21}$$

This ansatz, which is chosen as an approximation for a small elastic phase, is closely related to that by Forest (2013), Wulfinghoff et al. (2015), and Prahs and Böhlke (2019b) in the context of a two-phase laminate material. The constant macroscopic shear is denoted as $\bar{\gamma}$ and the periodic fluctuation as $\tilde{u}(x_1)$. The considered boundary value problem (BVP) is quasi-static. Body forces are neglected. The infinitesimal strain tensor ε and the plastic strain tensor ε^{p} read

$$
\begin{aligned}
\varepsilon_{\pm} &= \frac{1}{2}\left(\bar{\gamma} + \tilde{u}'_{\pm}(x_1)\right)\left(\boldsymbol{e}_1 \otimes \boldsymbol{e}_2 + \boldsymbol{e}_2 \otimes \boldsymbol{e}_1\right), \\
\varepsilon^{\mathrm{p}}_{\pm} &= \frac{1}{2}\gamma^{\pm}(x_1)\left(\boldsymbol{e}_1 \otimes \boldsymbol{e}_2 + \boldsymbol{e}_2 \otimes \boldsymbol{e}_1\right).
\end{aligned}
\tag{5.22}
$$

Here, $(\cdot)'$ denotes the derivative of a quantity with respect to x_1. Application of Hooke' law yields the Cauchy stress, given by

$$
\boldsymbol{\sigma}^{\pm} = G\left(\bar{\gamma} + \tilde{u}'_{\pm}(x_1) - \gamma^{\pm}(x_1)\right)\left(\boldsymbol{e}_1 \otimes \boldsymbol{e}_2 + \boldsymbol{e}_2 \otimes \boldsymbol{e}_1\right). \tag{5.23}
$$

Exploitation of the balance of linear momentum, cf. Eq. (3.12), for the quasi-static case and without body forces yields

$$
\tilde{u}''_{\pm}(x_1) = \gamma^{\pm\prime}(x_1), \quad \partial\gamma^{\pm}/\partial x_2 = 0. \tag{5.24}
$$

Equation $(5.24)_2$ states that the plastic slips $\gamma^{\pm}$ do not depend on x_2. Based on Eq. $(5.24)_1$, it is possible to give the following formulation for the fluctuation $\tilde{u}_{\pm}$ and its first derivative $\tilde{u}'_{\pm}$

$$
\begin{aligned}
\tilde{u}'_{\pm}(x_1) &= \gamma^{\pm}(x_1) + k^{\pm}, \\
\tilde{u}_{\pm}(x_1) &= \int \gamma^{\pm}(x_1)\,\mathrm{d}x_1 + x_1 k^{\pm} + d^{\pm}.
\end{aligned}
\tag{5.25}
$$

The integration constants that occur in this case are referred to as $k^{\pm}$ and $d^{\pm}$. Thus, the resolved shear stress reads

$$
\tau^{\pm} = G\left(\bar{\gamma} + \tilde{u}'_{\pm} - \gamma^{\pm}\right). \tag{5.26}
$$

Replacing the Laplacian $\Delta_{\mathrm{n}}\gamma^{\pm}$ by $\gamma^{\pm\prime\prime}$, Eq. (5.15) reduces to a an ODE with respect to x_1, reading

$$\gamma^{\pm\prime\prime} - \frac{\Theta_0}{K_g^{\pm}}\gamma^{\pm} = -\frac{G\left(\bar{\gamma}+k^{\pm}\right)}{K_g^{\pm}}. \tag{5.27}$$

The slip plane normals of the considered slip systems are orthogonal to the GB and, thus, $C = 1$ holds true. The corresponding boundary conditions according to Eq. (5.20) read

$$\begin{aligned} \gamma^{+\prime}(0) &= \frac{\kappa}{K_g^{+}}\left(\gamma^{+}(0)-\gamma^{-}(0)\right), \\ \gamma^{-\prime}(0) &= \frac{\kappa}{K_g^{-}}\left(\gamma^{+}(0)-\gamma^{-}(0)\right). \end{aligned} \tag{5.28}$$

Since the transition between the elastic phase and an elastoplastic phase is considered as micro-hard, $\gamma^{-}(h^{-}) = 0$ and $\gamma^{+}(h^{+}) = 0$ hold true. Moreover, Eq. (3.12) reduces to $\sigma_{12}^{\pm\prime} = 0$. Consequently, the stress $\boldsymbol{\sigma}$ is constant throughout the three-phase laminate material.

Analytic solution To determine the integration constants $k^{\pm}$ and $d^{\pm}$, four additional conditions are necessary. Here, $\tilde{u}(x_1)$ is considered to be a periodic fluctuation. Thus, the following two conditions have to be fulfilled

$$\int_{-h^-}^{h^++s} \tilde{u}(x_1)\,\mathrm{d}x_1 = \int_{-h^-}^{0} \tilde{u}_-(x_1)\,\mathrm{d}x_1 + \int_{0}^{h^++s} \tilde{u}_+(x_1)\,\mathrm{d}x_1 = 0, \tag{5.29}$$

$$\int_{-h^-}^{h^++s} \tilde{u}'(x_1)\,\mathrm{d}x_1 = \int_{-h^-}^{0} \tilde{u}'_-(x_1)\,\mathrm{d}x_1 + \int_{0}^{h^++s} \tilde{u}'_+(x_1)\,\mathrm{d}x_1 = 0, \tag{5.30}$$

cf. Wulfinghoff et al. (2015). Moreover, to ensure the continuity of the displacement field,

$$\tilde{u}_+(0) = \tilde{u}_-(0) \tag{5.31}$$

has to be satisfied. Regarding the quasi-static case and neglecting body forces, the balance of linear momentum at the GB, cf. Eq. (4.22), yields

$$\tilde{u}'_+(0) - \tilde{u}'_-(0) = \gamma^+(0) - \gamma^-(0). \tag{5.32}$$

Solving the ODE according to Eq. (5.27) along with the boundary conditions given in Eq. (5.28) yields the solutions of $\gamma^\pm(x_1)$ parameterized in $k^\pm$. The parameterized solutions are linear in $k^\pm$. Thus, according to Eq. (5.25), $\tilde{u}'(x_1)$ as well as $\tilde{u}(x_1)$ are linear in $k^\pm$ and in $d^\pm$. Consequently, Eqs.(5.29) to (5.32) constitute four equations, linear in $k^\pm$ and $d^\pm$, which can be solved by analytical means. As a consequence of Eq. (5.32), $k^+ = k^-$ holds true.

5.3.3 Influence of the defect parameter, the grain width, and the grain boundary strength on the solution

Effect of the defect parameter Subsequently the influence of a variation of K_g^+ on the distribution of the plastic slip and the fluctuation of the displacement field $\tilde{u}(x_1)$ is discussed. Regarding the left grain of the laminate, the defect parameter is chosen as $K_g^- = 84$ µN, cf. Bayerschen et al. (2015). The defect parameter K_g^+ is considered to be less than or equal to K_g^-. Both grains exhibit the same width. In this context, $h^- = h^+ = 0.5$ µm is considered. The width of the elastic phase is chosen as $s = h^+/4$. Regarding a macroscopic shear of $\bar{\gamma} = 0.01$, the distribution of the plastic slips $\gamma^-(x_1)$ and $\gamma^+(x_1)$ is depicted in Fig. 5.6. If both defect parameters coincide, the distribution of the plastic slip is continuous across the GB and the gradient of the plastic slips at the GB vanishes. Thus, $\gamma^+(0) = \gamma^-(0)$ and $\gamma^{+\prime}(0) = \gamma^{-\prime}(0) = 0$ hold true. Regarding the case $K_g^+ < K_g^-$, the distribution of the plastic slip becomes discontinuous at the GB. According to $\gamma^{+\prime}(0)/\gamma^{-\prime}(0) = K_g^-/K_g^+$, cf. Eq. (5.28), the smaller the value of K_g^+, the higher the absolute value of the gradient of γ^+ at the GB. Thus, the GB, which is of artificial nature for coinciding defect parameters, becomes clearly visible for different values of K_g^+ and K_g^-. Moreover, the absolute value of $\gamma^{+\prime}(0)$ is significantly bigger than $\gamma^{-\prime}(0)$ for a decreasing value of K_g^+. The defect parameter $K_g^\pm$ is inversely proportional to the initial dislocation density $\rho_0^\pm$, cf. Bayerschen et al. (2016b). Thus, the case $K_g^+ < K_g^-$ implies that the initial dislocation density in the right grain is higher than in the left grain, i.e., $\rho_0^+ > \rho_0^-$. Consequently the pile-up of dislocations in the right grain is expected to be more pronounced than in the left grain. This expectation is met by the model at hand. A high gradient of the plastic slip reflects a big dislocation pile-up at the GB, cf. Bayerschen and Böhlke (2016).

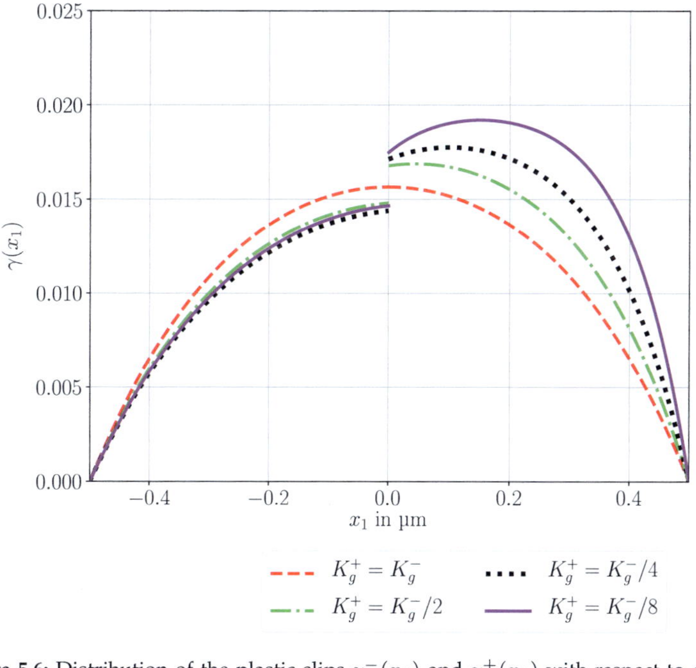

Figure 5.6: Distribution of the plastic slips $\gamma^-(x_1)$ and $\gamma^+(x_1)$ with respect to different values of K_g^+. The distribution is continuous if the defect parameters K_g^- and K_g^+ coincide. The absolute value of $\gamma^{+\prime}(0)$ increases with decreasing K_g^+, indicating a bigger dislocation pile-up in the right grain.

Figure 5.7 depicts the distribution of $\tilde{u}(x_1)$ for different values of the defect parameter. It is continuously differentiable at the GB in case of coinciding defect parameters. The absolute value of $\tilde{u}(0)$ is zero. If $K_g^+ < K_g^-$ is considered, the distribution of $\tilde{u}(x_1)$ is not continuously differentiable at the GB. Hence, the jump of K_g at the GB induces a jump of the total strain. However, the Hadamard condition, cf. (Gurtin et al., 2010, p. 210), is fulfilled, cf. Eq. (A.16). The decrease in K_g^+ shifts $\tilde{u}(0)$ toward negative values. Moreover, the amplitude of $\tilde{u}(x_1)$ increases with a decrease in K_g^+.

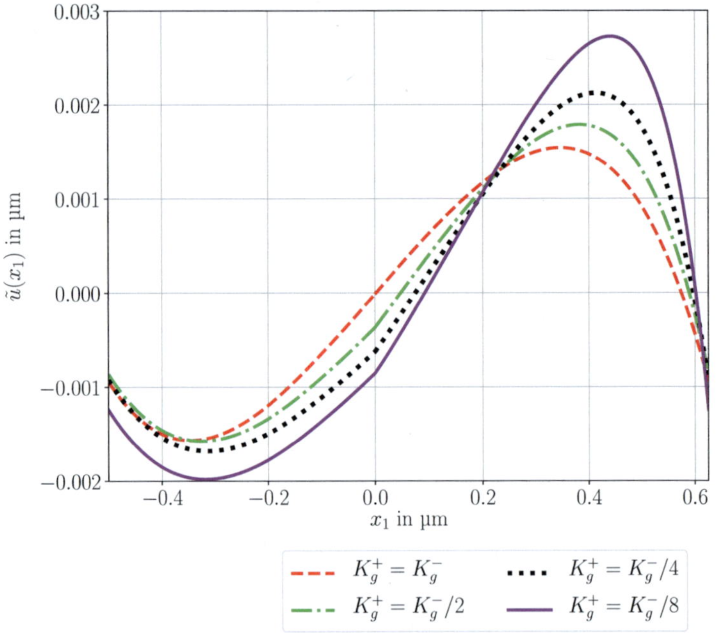

Figure 5.7: Distribution of the fluctuation $\tilde{u}(x_1)$ with respect to different values of K_g^+. The distribution is continuously differentiable at the GB, if both defect parameters K_g^+ and K_g^- coincide. If the parameters differ in their value, the distribution exhibits a kink at the GB. The amplitude of $\tilde{u}(x_1)$ increases if K_g^+ decreases.

Effect of the relative width of the plastic phases The influence of a variation of the width of the right grain on the distribution of the plastic slips and the fluctuation $\tilde{u}(x_1)$ is presented, in the following. To this end, the defect parameters are chosen equal, with the amplitude $K_g^{\pm} = 84$ µN. The width of the left grain is considered as $h^- = 0.5$ µm, and the width of the elastic phase as $s = h^-/4$. The width h^+ of the right grain is chosen less than or equal to h^-. The distribution of the plastic slips $\gamma^-(x_1)$ and $\gamma^+(x_1)$ is depicted for a macroscopic shear of $\bar{\gamma} = 0.01$ in Fig. 5.8.

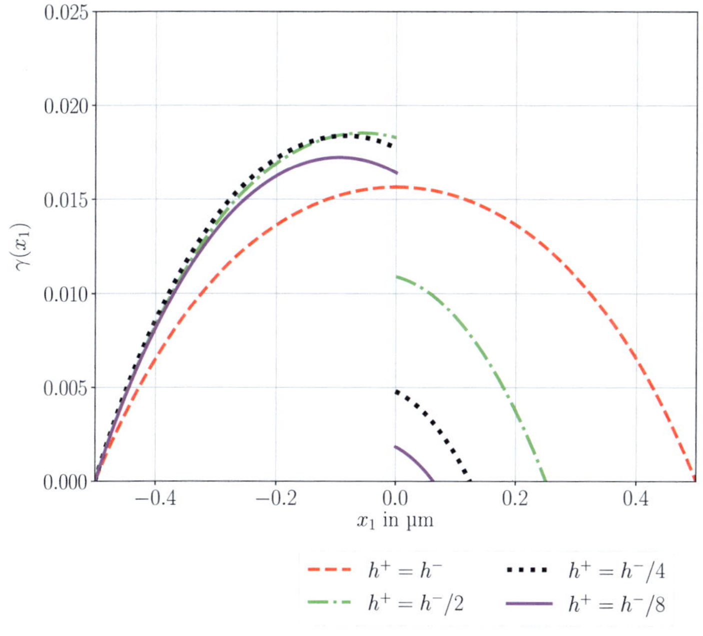

Figure 5.8: Distribution of the plastic slips $\gamma^-(x_1)$ and $\gamma^+(x_1)$ with respect to different values of h^+. The width of the right grain is indicated by a vanishing value of the respective $\gamma^+(x_1)$ distribution. The defect parameters coincide, i.e., $K_g^+ = K_g^-$. The distribution is continuous across the GB if the width of both grains coincide. The gradients of the plastic slips at the GB coincide, irrespective of the choice of h^+, i.e., $\gamma^{-\prime}(0) = \gamma^{+\prime}(0)$ holds true.

If both grains are of the same width, i.e., $h^- = h^+$, the distribution of the plastic slips is continuous across the GB and the corresponding gradients are vanishing, i.e., $\gamma^-(0) = \gamma^+(0)$ and $\gamma^{-\prime}(0) = \gamma^{+\prime}(0) = 0$. Regarding $h^+ < h^-$, the distribution of the plastic slips exhibits a jump at the GB. This behavior is somehow unphysical. The distribution of the plastic slip as well as its gradient are expected to be continuous if the same material parameters and coherent lattice orientations are considered. Here, only the value of the gradients of the plastic slip coincide at the GB. Subsequently, it is shown that the jump of the plastic slip, induced by the variation of the width of the right grain, can be compensated by an appropriate choice of the defect parameters in both grains.

The distribution of $\tilde{u}(x_1)$, cf. Fig. 5.9, is continuously differentiable across the GB, if $h^+ = h^-$ holds true. Consequently, the jump of the width of both grains induces a jump of the total strain at the GB. Nevertheless, the Hadamard condition is fulfilled, cf. Eq. (A.16). Regarding a decrease in h^+ leads to a kink of the distribution at the GB and a shift of $\tilde{u}(0)$ towards positive values.

Consistency and continuous differentiability at the GB Both the variation of K_g^+ while holding K_g^- fixed, or the variation of h^+ for a fixed value of h^-, yields discontinuities of the plastic slip and the total strain at the GB. This motivates the question if it is possible to find a relation between the defect parameters, K_g^+ and K_g^-, and the widths, h^+ and h^-, of both grains such that both the plastic slip and the total strain are continuous across the GB. Such a relationship is provided by the analytical solution of the plastic slip in case of the laminate material under consideration. Regarding the case $h^+ = \alpha h^-$, the specific choice $K_g^+ = \alpha^2 K_g^-$ yields a continuous distribution of γ and a continuously differentiable distribution of $\tilde{u}$ across the GB, as depicted in Fig. 5.10 and Fig. 5.11. Thus, consistency

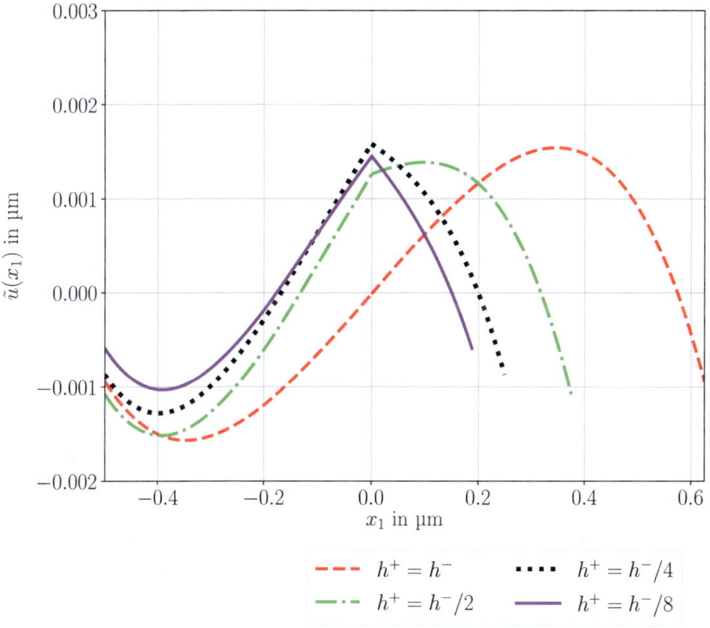

Figure 5.9: Distribution of the fluctuation $\tilde{u}(x_1)$ with respect to different values of h^+. The defect parameters coincide, i.e., $K_g^+ = K_g^-$. The distribution is continuously differentiable at the GB if both grains are of the same width. If the width h^+ decreases, the distribution exhibits a kink at the GB.

of the plastic slip and the total strain is obtained, if

$$\frac{K_g^+}{K_g^-} = \left(\frac{h^+}{h^-}\right)^2 \tag{5.33}$$

holds true. Taking into account $K_g^\pm = E/\rho_0^\pm$, cf. Bayerschen et al. (2016b), Eq. (5.33) can be formulated in terms of the initial dislocation densities of the corresponding grain, reading

$$\frac{\rho_0^-}{\rho_0^+} = \left(\frac{h^+}{h^-}\right)^2 \tag{5.34}$$

Consequently, holding ρ_0^- and h^- fixed, the initial dislocation density ρ_0^+ increases quadratically for a decrease of h^+ in order to guarantee the continuity of the distributions of the plastic slip and the total strain at the GB.

Effect of the grain boundary strength Subsequently, the influence of the grain boundary strength κ on the distribution of the plastic slip and the fluctuation $\tilde{u}(x_1)$ is illustrated in Fig. 5.12 and Fig. 5.13. The defect parameters are chosen as $K_g^- = 84$ µN and $K_g^+ = K_g^-/2$. The width of both grains is $h^+ = h^- = 0.5$ µm. The grain boundary strength κ is varied in the interval $[10^0\ \mathrm{N/m}, 10^6\ \mathrm{N/m}]$. For small values of κ, a distinct jump of the plastic slip is obtained at the transition between both elastoplastic phases. In contrast, high values of κ yield to a decrease of the jump of the plastic slip at the transition.

Compared to small values of κ, increasing values lead to a higher smootheness of the distribution $\tilde{u}(x_1)$ at the transition. Contrarily, small values of κ yield a jump of the strain at the GB. The overall shape of the fluctuation $\tilde{u}(x_1)$ is nearly not affected by the choice of the grain boundary strength κ.

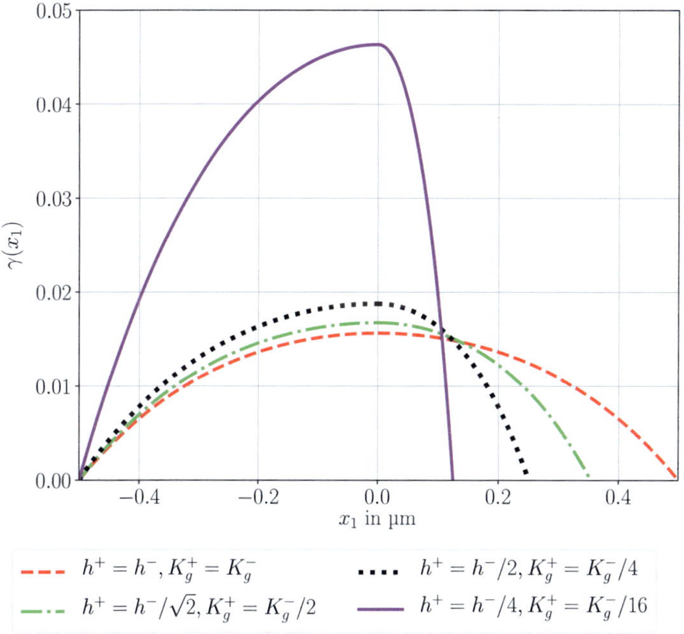

Figure 5.10: Distribution of the plastic slips $\gamma^-(x_1)$ and $\gamma^+(x_1)$ with respect to different values of h^+ and K_g^+. While h^- and K_g^- are fixed, $h^+ = \alpha h^-$ and $K_g^+ = \alpha^2 K_g^-$ are considered with $\alpha \in \{1, 1/\sqrt{2}, 1/2, 1/4\}$, $K_g^- = 84$ µN, $h^- = 0.5$ µm and $s = h^-/4$. For the specific choice of K_g^+ and h^+, the distribution of the plastic slips is continuous across the GB.

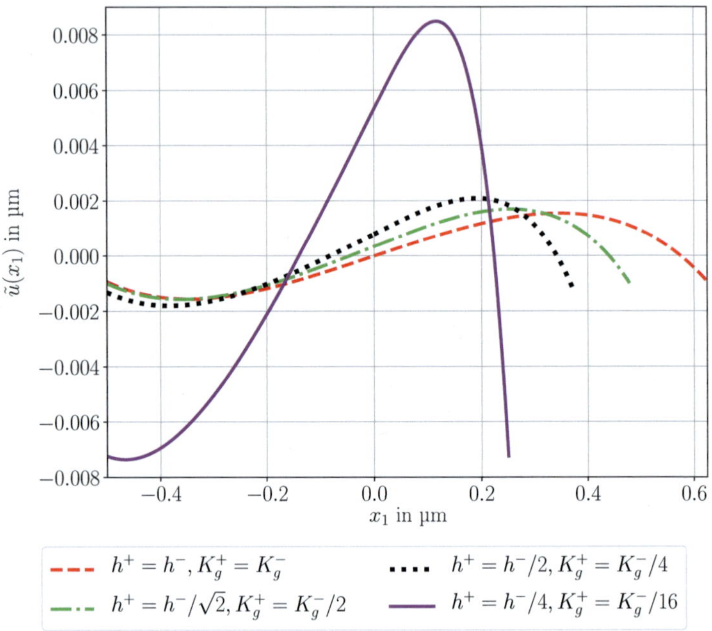

Figure 5.11: Distribution of the fluctuation $\tilde{u}$ with respect to different values of h^+ and K_g^+. While h^- and K_g^- are fixed, $h^+ = \alpha h^-$ and $K_g^+ = \alpha^2 K_g^-$ are considered with $\alpha \in \{1, 1/\sqrt{2}, 1/2, 1/4\}$, $K_g^- = 84$ µN, $h^- = 0.5$ µm and $s = h^-/4$. For the specific choice of K_g^+ and h^+, the distribution of $\tilde{u}$ is continuously differentiable across the GB. Thus, the distribution of the total strain is continuous across the GB.

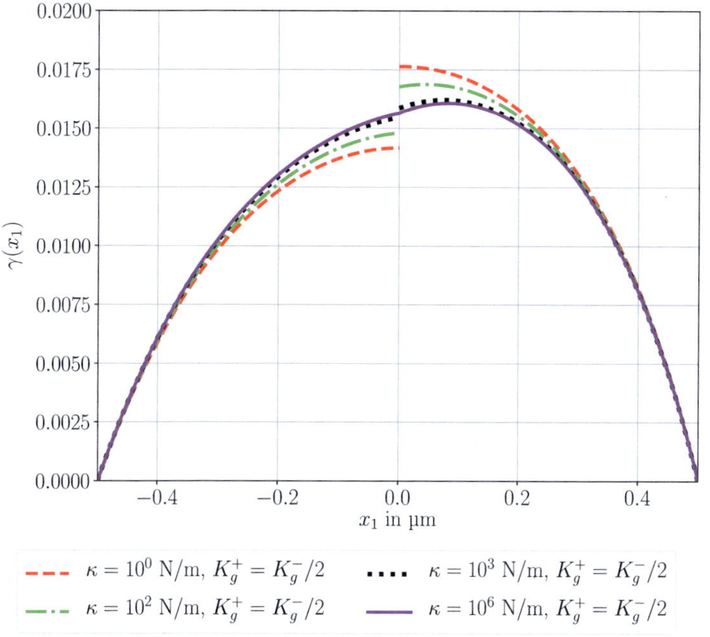

Figure 5.12: Choosing a high value of κ leads to a continuous distribution of the plastic slip at the transition between both elastoplastic phases. Small values of κ lead to a distinct jump of the plastic slip.

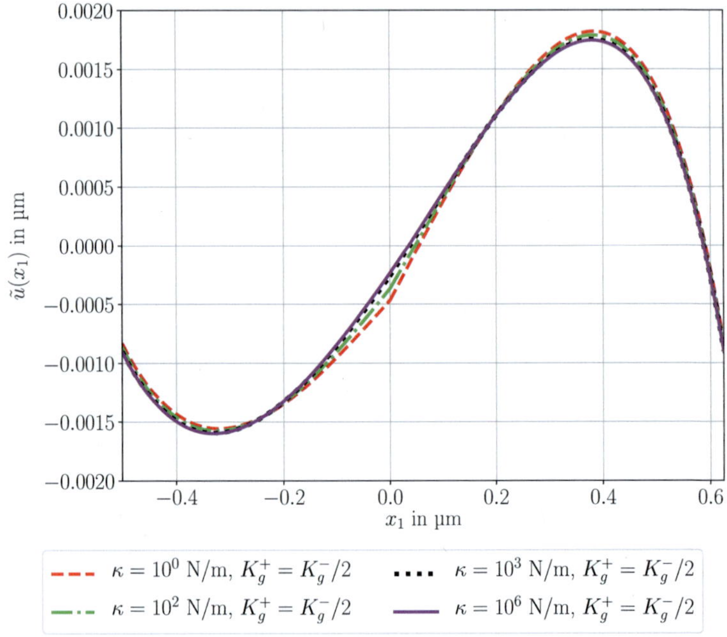

Figure 5.13: A high value of κ yields a smooth distribution of the fluctuation $\tilde{u}(x_1)$ at the transition between both elastoplastic phases. Small values of κ leads to a kink in the distribution of $\tilde{u}$, and, thus, to a jump of the strain at the transition. The choice of κ shows nearly no influence on the shape of the fluctuation $\tilde{u}(x_1)$.

5.4 Interim Conclusions

Regarding the two- and three-phase laminate in the context of a slip gradient crystal plasticity theory, the obtained results are summarized.

Two-phase laminate

- A vanishing defect parameter yields a constant distribution of the plastic slip without any gradients. This constitutes the upper bound for the plastic slip.
- Regarding small values of the defect parameter, the distribution of the plastic slip exhibits a plateau-like behavior.
- The higher the defect parameter the more parabolic the distribution of the plastic slip.
- If the initial hardening modulus is neglected, a parabolic distribution of the plastic slip is obtained as upper bound.
- Increasing the initial hardening modulus decreases the absolut value of the plastic slip and its gradient at the transition to the elastic phase.

Three-phase laminate

- Regarding the two elastoplastic phases of the considered three-phase laminate material, GB effects are present due to different values of the defect parameter in each grain. This effect is present even though coherent slip systems are considered within the elastoplastic phases.
- The variation of the width of one elastoplastic phase leads to a discontinuity of the distribution of the plastic slip at the GB. This somehow nonphysical behavior of the plastic slips at the GB, i.e., the inconsistency in the coherent case, is based on the assumption that the slip rates are independent. Alternatively, the special case of a continuous plastic slip and its gradient in case of a coherent crystal structure could be obtained by adding additional slip transfer criteria.
- A specific choice of the defect parameters of both elastoplastic phases with respect to their widths yields a continuous distribution of the

plastic slip. Regarding the same defect parameter for both elastoplastic phases, the absolute value of the gradients of the plastic slip at the GB coincide.

- High values of the grain boundary strength yield a continuous distribution of the plastic slip at the GB.

Chapter 6

Summary and Conclusions

In the context of micro-structured materials, extended continua are applied to account for size effects and nonlocal mechanical behavior. Regarding crystals such as metals, slip gradient crystal plasticity theories yield results that are in good agreement with experiments. To obtain such an enriched theory, several methods are available from literature to extend the classical Cauchy continuum theory. Especially the application of an extended PoVP is widely spread. It offers a straight-forward derivation of additional field equations. However, the PoVP also exhibits redundancies if thermodynamically consistent constitutive equations are to be derived and is, consequently, not a self-consistent method to obtain a closed theory. Moreover, singular surfaces such as GBs are commonly considered in a quasi-static sense. The curvature of such material singular surfaces is not accounted for intrinsically. In the work at hand it is shown that the results obtained by invariance considerations of an extended energy balance are equivalent to an extended PoVP. In contrast to an extended PoVP, the invariance of an extended energy balance naturally provides an extended balance of internal energy. In addition, the geometrical properties of a material singular surface are also automatically taken into account. However, the derivation of the necessary field equations is more involved than the application of an extended PoVP. Subsequently, the main results of the work at hand are summarized and concluding remarks are given for each chapter separately.

Review on extended continua

- Regarding the presented methods, only the invariance of an extended energy balance or the application of an extended PoVP yields results that are not restricted to a specific class of materials.
- To obtain thermodynamically consistent constitutive equations, an extended PoVP requires the statement of an extended balance of internal energy, in addition.
- The invariance of an extended balance of energy automatically provides the extended balance of internal energy. However, no additional balance equations are obtained.
- Thus, the question concerning the relationship between the balance equations obtained by an extended PoVP and an extended energy balance is motivated.

Regarding the mentioned methods, only the invariance considerations of an extended energy balance exhibits self-consistency with respect to the derivation of a closed theoretical framework of balance laws and corresponding constitutive equations.

On invariance properties of an extended energy balance

- The consideration of an additional DOF does not provide an associated, additional force balance.
- Regarding a scalar or a vectorial additional DOF, the conservation of mass as well as the balance of linear momentum are obtained in the same form as for a classical continuum. However, only with an additional scalar DOF, the Cauchy stress tenor remains symmetric.
- The exploitation of the Clausius-Duhem inequality leads to a thermodynamically consistent, nonlocal flow rule for the scalar-valued DOF. Partitioning of the obtained flow rule yields the so-called micro force balance as a constitutive equation. For an isothermal material behavior with a homogeneous temperature distribution and negligible effects

due to heat supply, the nonlocal flow rule for the additional DOF is reduced to a PDE.

- The equivalence between the extended energy balance and an extended PoVP is shown for a scalar-valued DOF.

It is not possible to obtain an additional balance equation related to a supplementary DOF. However, the set of equations defining the theoretical framework are equivalent for both an extended PoVP and an extended energy balance. Starting with an extended energy balance, the so-called micro-force balance is merely a PDE obtained by the partitioning of the nonlocal flow rule. In this context, the notion of balance is misleading for this PDE.

On interface conditions on a material singular surface

- The balance of mass, linear momentum and internal energy are obtained by invariance considerations. The balance of mass explicitly considers the mean curvature of the GB. Both the balance of linear momentum and internal energy depend on the curvature of the GB only via the balance of mass.
- As a result of the balance equations, obtained by invariance considerations, and the exploitation of the Coleman-Noll procedure, flow rules for the plastic slip at the GB are obtained.
- The GB behaves purely elastically if the material behavior is considered to be isothermal with a homogeneous temperature distribution and if effects due to heat supply are omitted.
- It is shown that the framework discussed is equivalent to an extended PoVP.

The mean curvature of the GB is directly accounted for by the conservation of mass at the GB. This result is automatically obtained in the context of an extended energy balance, since the transport theorem for singular surfaces is directly applied. Several flow rules are admissible in general. However, the GB behaves purely elastic in the context of an isothermal material behavior with homogeneous temperature distribution and negligible effects due to heat supply. This special case is considered by the majority of gradient crystal plasticity theories in literature. Moreover, the invariance considerations of an extended energy balance yield the same set of equations obtained by an extended principle of virtual power.

Application to a slip gradient crystal plasticity In this context, a small strain framework is considered with respect to single slip. Both a two-phase and a three-phase elastoplastic laminate material is considered. The periodic two-phase laminate material periodically consists of alternating elastoplastic and elastic phases. Thereby, the elastoplastic phase mimicks the behavior of a single crystal. The three-phase laminate material consists of two adjacent elastoplastic phases and a subsequent elastic phase. Thus, the two elastoplastic phases reflect the behavior of a bicrystal. For brevity, the slip systems in both grains are considered geometrically coherent. With this setup, the following results are obtained.

- An upper bound for the plastic slip is obtained by either the choice of a vanishing defect parameter, or a vanishing initial hardening modulus.
- Regarding two coherent slip systems of a bicrystal, GB effects occur based on different defect parameters, and, thus, different initial dislocation densities in each grain.
- At the GB, discontinuities of the plastic slip are obtained by a jump of the defect parameter or the grain width. A continuous distribution of the plastic slip results from a specific choice of the grain widths and the defect parameter.
- The GB strength significantly influences the behavior of the plastic slip at the GB. A continuous distribution can be achieved for high values of the GB strength.

The obtained distribution of the plastic slip is highly sensitive to a change of the material parameters. The variation of the defect parameter indicates, that GB effects are not only present at actual GBs separating two grains with different orientation from each other. Slip gradients and discontinuities of the plastic slip are also observed at the transition of regions that differ solely with respect to the corresponding initial dislocation density.

Outlook Slip gradient crystal plasticity theories constitute coarse grained models with respect to the prediction of dislocation based plasticity. Instead, the dislocation density tensor, i.e., the curl of the plastic distortion can be considered as additional DOF. This yields a continuum mechanical model which is closely related to the physical mechanisms of plasticity. To this end, the invariance properties of an energy balance extended by a second-order tensor should be considered.

Appendix A

Appendix

A.1 Equivalence of an Extended Principle of Virtual Power and an Extended Energy Balance

Micro-force balance The exploitation of an extended principle of virtual power yields the so-called micro-force balance, reading

$$a^l = \operatorname{div}\left(\boldsymbol{b}^l\right), \tag{A.1}$$

cf. Forest (2009, Eq. (32)).

Balance of internal energy Accounting for l according to Eq. (2.59), the balance of internal energy is given by

$$\rho \dot{e} = p^{(i)} - \operatorname{div}(\boldsymbol{q}) + \rho r, \quad p^{(i)} = \boldsymbol{\sigma} \cdot \dot{\boldsymbol{\varepsilon}} + a^l\, \delta\dot{\Phi} + \boldsymbol{b}^l \cdot \operatorname{grad}\left(\delta\dot{\Phi}\right), \tag{A.2}$$

cf. Forest (2009, Eq. (13)).

Clausius-Duhem inequality regarding an additional scalar-valued DOF The dissipation inequality in terms of the Clausius-Duhem inequality can be written as

$$-\rho\dot{\psi} - \rho\eta\dot{\theta} + p^{(i)} - \frac{1}{\theta}\boldsymbol{q} \cdot \boldsymbol{g} \geq 0, \tag{A.3}$$

cf. Forest (2009, Eq. (15)). Similar to Eq. (3.28), the following decomposition of the specific free energy is considered

$$\psi(\boldsymbol{\varepsilon}-\boldsymbol{\varepsilon}^{\mathrm{p}},\phi,\nabla\phi,\theta)=\psi_{\mathrm{e}}(\boldsymbol{\varepsilon}-\boldsymbol{\varepsilon}^{\mathrm{p}})+\psi_{\mathrm{h}}(\phi)+\psi_{\mathrm{g}}(\nabla\phi)+\psi_{\theta}(\theta). \tag{A.4}$$

Thus, the Clausius-Duhem inequality takes the following form

$$\begin{aligned}\rho\delta=&\left(\boldsymbol{\sigma}-\rho\frac{\partial\psi_{\mathrm{e}}}{\partial\boldsymbol{\varepsilon}}\right)\cdot\dot{\boldsymbol{\varepsilon}}-\rho\frac{\partial\psi_{\mathrm{e}}}{\partial\boldsymbol{\varepsilon}^{\mathrm{p}}}\cdot\frac{\partial\boldsymbol{\varepsilon}^{\mathrm{p}}}{\partial\phi}\dot{\phi}-\rho\left(\eta+\frac{\partial\psi_{\theta}}{\partial\theta}\right)\dot{\theta}-\boldsymbol{q}\cdot\boldsymbol{g}/\theta\\&+\left(a^{l}-\rho\frac{\partial\psi_{\mathrm{h}}}{\partial\phi}\right)\dot{\phi}+\left(\boldsymbol{b}^{l}-\rho\frac{\partial\psi_{\mathrm{g}}}{\partial\nabla\phi}\right)\cdot\nabla\dot{\phi}\geq0.\end{aligned} \tag{A.5}$$

This yields the potential relations, similar to Eq. (3.30), reading

$$\boldsymbol{\sigma}=\rho\frac{\partial\psi_{\mathrm{e}}}{\partial\boldsymbol{\varepsilon}},\quad\eta=-\frac{\partial\psi_{\theta}}{\partial\theta},\quad\boldsymbol{b}^{l}=\rho\frac{\partial\psi_{\mathrm{g}}}{\partial\nabla\phi}. \tag{A.6}$$

Consequently, the reduced dissipation inequality is given by

$$\left(a^{l}-\rho\frac{\partial\psi_{\mathrm{h}}}{\partial\phi}-\rho\frac{\partial\psi_{\mathrm{e}}}{\partial\boldsymbol{\varepsilon}^{\mathrm{p}}}\cdot\frac{\partial\boldsymbol{\varepsilon}^{\mathrm{p}}}{\partial\phi}\right)\dot{\phi}-\boldsymbol{q}\cdot\boldsymbol{g}/\theta\geq0. \tag{A.7}$$

The mechanical dissipation is represented by the first term, the thermal dissipation by the second expression.

Flow rule Accounting for Fourier's law, the positivity of the thermal dissipation is fulfilled. Finally, the mechanical dissipation must not be negative, i.e., a proper choice for $\dot{\phi}$ is necessary. Regarding linear irreversible thermodynamics, for brevity, $\dot{\phi}$ reads

$$\dot{\phi}=\dot{\phi}_0\left(a^{l}-\rho\frac{\partial\psi_{\mathrm{h}}}{\partial\phi}-\rho\frac{\partial\psi_{\mathrm{e}}}{\partial\boldsymbol{\varepsilon}^{\mathrm{p}}}\cdot\frac{\partial\boldsymbol{\varepsilon}^{\mathrm{p}}}{\partial\phi}\right),\quad\dot{\phi}_0\geq0, \tag{A.8}$$

with a constant, referential rate $\dot{\phi}_0$. The flow rule according to Eq. (A.8) does not contain a divergence term. This ODE is referred to as local flow rule.

Comparison to the presented theory Taking into account Eq. (A.1), the local flow rule according to Eq. (A.8) reads

$$\dot{\phi} = \dot{\phi}_0 \left(\operatorname{div}\left(\boldsymbol{b}^l\right) - \rho \frac{\partial \psi_\mathrm{h}}{\partial \phi} - \rho \frac{\partial \psi_\mathrm{e}}{\partial \boldsymbol{\varepsilon}^\mathrm{p}} \cdot \frac{\partial \boldsymbol{\varepsilon}^\mathrm{p}}{\partial \phi} \right), \quad \dot{\phi}_0 \geq 0. \tag{A.9}$$

Thus, the combination of the so-called micro-force balance and the flow rule according to Eq. (A.8) is equivalent to the nonlocal flow rule according to Eq. (3.32). Moreover, combining Eq. (A.7) and Eq. (A.1) is equivalent to Eq. (3.31). This result is independent of the considered flow rule. Consequently, the extended principle of virtual power and the extended energy balance yield the same set of equations and, thus, are equivalent.

A.2 Application to a Slip Gradient Crystal Plasticity

A.2.1 Two-phase laminate material

Integration constant Integration of Eq. (5.7) twice with respect to x_1 yields

$$\tilde{u}(x_1) = \int \gamma(x_1)\,\mathrm{d}x_1 + c x_1 + d. \tag{A.10}$$

The integration constant c is determined by evaluating Eq. $(5.9)_2$, reading

$$\begin{aligned}
c &= \frac{Z_1 + Z_2}{N_1 + N_2 + N_3},\\
Z_1 &= -2\left(\left(-1/2 + (h+s/2)\sqrt{a}\right) \mathrm{e}^{\sqrt{a}(2\,h+s)} - 1/2\,\mathrm{e}^{2\,(h+s)\sqrt{a}}\right) G\bar{\gamma},\\
Z_2 &= -2\left(1/2 + \left(1/2 + (h+s/2)\sqrt{a}\right) \mathrm{e}^{\sqrt{a}s}\right) G\bar{\gamma},\\
N_1 &= \left(2\,K_g\,(h+s/2)\,a^{3/2} + 2\,\left(-1/2 + (h+s/2)\sqrt{a}\right) G\right) \mathrm{e}^{\sqrt{a}(2\,h+s)},\\
N_2 &= \left(2\,K_g\,(h+s/2)\,a^{3/2} + 2\,G\left(1/2 + (h+s/2)\sqrt{a}\right)\right) \mathrm{e}^{\sqrt{a}s},\\
N_3 &= -G\mathrm{e}^{2\,(h+s)\sqrt{a}} + G. \qquad \text{(A.11)}
\end{aligned}$$

In this context, the abbreviation $a = \Theta_0/K_g$ is introduced. Moreover, evaluation of Eq. $(5.9)_1$ yields the integration constant d as

$$\begin{aligned}
d &= \frac{Z_3 + Z_4}{N_4 + N_5 + N_6},\\
Z_3 &= -\left(-\mathrm{e}^{(2\,h+s)\sqrt{a}}s\sqrt{a} - \mathrm{e}^{(2\,s+2\,h)\sqrt{a}}\sqrt{a}s + \mathrm{e}^{\sqrt{a}s}s\sqrt{a}\right) G\bar{\gamma},\\
Z_4 &= -\left(\sqrt{a}s - 2\,\mathrm{e}^{(2\,h+s)\sqrt{a}} + 2\,\mathrm{e}^{(2\,s+2\,h)\sqrt{a}} - 2\,\mathrm{e}^{\sqrt{a}s} + 2\right) G\bar{\gamma},\\
N_4 &= 2\sqrt{a}G\mathrm{e}^{(2\,s+2\,h)\sqrt{a}} - 2\,G\sqrt{a},\\
N_5 &= \left(2\,G\sqrt{a} + (-4\,h - 2\,s)\,aG + (-4\,h - 2\,s)\,K_g\,a^2\right) \mathrm{e}^{(2\,h+s)\sqrt{a}},\\
N_6 &= \left(-2\,G\sqrt{a} + (-4\,h - 2\,s)\,aG + (-4\,h - 2\,s)\,K_g\,a^2\right) \mathrm{e}^{\sqrt{a}s}. \qquad \text{(A.12)}
\end{aligned}$$

A.2.2 Three-phase laminate material

Fulfillment of the Hadamard-condition Regarding the quasi-static case, the Hadamard condition is given by

$$[\![\boldsymbol{F}]\!]\boldsymbol{P} = \boldsymbol{0}, \quad \boldsymbol{P} = \boldsymbol{I} - \boldsymbol{n}^S \otimes \boldsymbol{n}^S, \qquad \text{(A.13)}$$

cf. Gurtin et al. (2010, pp. 209, 210). Accounting for Eq. (5.21), $\boldsymbol{F}$ reads

$$\boldsymbol{F} = -1\,(\boldsymbol{e}_1 \otimes \boldsymbol{e}_1 + \boldsymbol{e}_2 \otimes \boldsymbol{e}_2 + \boldsymbol{e}_3 \otimes \boldsymbol{e}_3) + \bar{\gamma}\boldsymbol{e}_1 \otimes \boldsymbol{e}_2 + \tilde{u}'\boldsymbol{e}_2 \otimes \boldsymbol{e}_1. \quad \text{(A.14)}$$

With respect to the three-phase laminate material, $\boldsymbol{n}^S = \boldsymbol{e}_1$ holds true. Thus, $\{\!\!\{\boldsymbol{F}\}\!\!\}$ and $\boldsymbol{P}$ from Eq. (A.13) read

$$\{\!\!\{\boldsymbol{F}\}\!\!\} = \left(\tilde{u}'_+ - \tilde{u}'_-\right)(\boldsymbol{e}_2 \otimes \boldsymbol{e}_1), \quad \boldsymbol{P} = \boldsymbol{e}_2 \otimes \boldsymbol{e}_2 + \boldsymbol{e}_3 \otimes \boldsymbol{e}_3. \quad \text{(A.15)}$$

Consequently, Eq. $(\text{A.13})_1$ is fulfilled according to

$$\{\!\!\{\boldsymbol{F}\}\!\!\}\boldsymbol{P} = \left(\tilde{u}'_+ - \tilde{u}'_-\right)((\boldsymbol{e}_1 \cdot \boldsymbol{e}_2)\,\boldsymbol{e}_2 \otimes \boldsymbol{e}_2 + (\boldsymbol{e}_1 \cdot \boldsymbol{e}_3)\,\boldsymbol{e}_2 \otimes \boldsymbol{e}_3) \equiv \boldsymbol{0}. \quad \text{(A.16)}$$

Bibliography

Agiasofitou, E., Lazar, M., 2009. Conservation and balance laws in linear elasticity of grade three. Journal of Elasticity 94 (1), 69–85.

Aifantis, K., Senger, J., Weygand, D., Zaiser, M., 2009. Discrete dislocation dynamics simulation and continuum modeling of plastic boundary layers in tricrystal micropillars. IOP Conference Series: Materials Science and Engineering 3 (1), 1–6.

Aifantis, K., Soer, W., Hosson, J. D., Willis, J., 2006. Interfaces within strain gradient plasticity: Theory and experiments. Acta Materialia 54 (19), 5077 – 5085.

Aifantis, K., Willis, J., 2005. The role of interfaces in enhancing the yield strength of composites and polycrystals. Journal of the Mechanics and Physics of Solids 53 (5), 1047 – 1070.

Armstrong, R., Codd, I., Douthwaite, R. M., Petch, N. J., 1962. The plastic deformation of polycrystalline aggregates. Philosophical Magazine 7 (73), 45–58.

Ashby, M. F., 1970. The deformation of plastically non-homogeneous materials. Philosophical Magazine 21 (170), 399–424.

Auffray, N., dell'Isola, F., Eremeyev, V., Madeo, A., Rosi, G., 2015. Analytical continuum mechanics à la Hamilton–Piola least action principle for second gradient continua and capillary fluids. Mathematics and Mechanics of Solids 20 (4), 375–417.

Bardella, L., 2010. Size effects in phenomenological strain gradient plasticity constitutively involving the plastic spin. International Journal of Engineering Science 48 (5), 550–568.

Bayerschen, E., 2017. Single-crystal gradient plasticity with an accumulated plastic slip: Theory and applications. Doctoral Dissertation, KIT Scientific Publishing, Schriftenreihe Kontinuumsmechanik im Maschinenbau Nr. 9, Karlsruhe.

Bayerschen, E., Böhlke, T., 2016. Power-law defect energy in a single-crystal gradient plasticity framework: a computational study. Computational Mechanics 58 (1), 13–27.

Bayerschen, E., McBride, A. T., Reddy, B. D., Böhlke, T., 2016a. Review on slip transmission criteria in experiments and crystal plasticity models. Journal of Materials Science 51 (5), 2243–2258.

Bayerschen, E., Prahs, A., Wulfinghoff, S., Ziemann, M., Gruber, P. A., Walter, M., Böhlke, T., 2016b. Modeling contrary size effects of tensile- and torsion-loaded oligocrystalline gold microwires. Journal of Materials Science 51 (16), 7451–7470.

Bayerschen, E., Stricker, M., Wulfinghoff, S., Weygand, D., Böhlke, T., 2015. Equivalent plastic strain gradient plasticity with grain boundary hardening and comparison to discrete dislocation dynamics. Proceedings of the Royal Society A 471, 1–19.

Beegle, B. L., Modell, M., Reid, R. C., 1974. Legendre transforms and their application in thermodynamics. AIChE Journal 20 (6), 1194–1200.

Bertram, A., 2005. Elasticity and Plasticity of Large Deformations: An Introduction. Springer, Berlin.

Bertram, A., 2015. Solid Mechanics: Theory, Modeling, and Problems. Springer, Heidelberg, New York, London.

Bertram, A., Forest, S., 2007. Mechanics based on an objective power functional. Technische Mechanik 27, 1–17.

Bertram, A., Krawietz, A., 2012. On the introduction of thermoplasticity. Acta Mechanica 223 (10), 2257–2268.

Bilby, B., Bullough, R., Smith, E., 1955. Continuous distributions of dislocations: a new application of the methods of non-Riemannian geometry. Proceedings of the Royal Society A 231 (1185), 263–273.

Capriz, G., 1989. Continua with Microstructure. Springer, New York.

Capriz, G., Mariano, P. M., 2003. Symmetries and Hamiltonian formalism for complex materials. Journal of Elasticity 72 (1-3), 57–70.

Capriz, G., Podio-Guidugli, P., Williams, W., 1982. On balance equations for materials with affine structure. Meccanica 17 (2), 80–84.

Cermelli, P., Fried, E., Gurtin, M. E., 2005. Transport relations for surface integrals arising in the formulation of balance laws for evolving fluid interfaces. Journal of Fluid Mechanics 544, 339–351.

Cermelli, P., Gurtin, M. E., 2002. Geometrically necessary dislocations in viscoplastic single crystals and bicrystals undergoing small deformations. International Journal of Solids and Structures 39 (26), 6281–6309.

Coleman, B. D., Gurtin, M. E., 1967. Thermodynamics with internal state variables. Journal of Chemical Physics 47 (2), 597–613.

Coleman, B. D., Noll, W., 1963. The thermodynamics of elastic materials with heat conduction and viscosity. Archive for Rational Mechanics and Analysis 13 (1), 167–178.

Cosserat, E., Cosserat, F., 1909. Théorie des Corps Déformables. Hermann, Paris.

dell'Isola, F., Madeo, A., Seppecher, P., 2016. Cauchy tetrahedron argument applied to higher contact interactions. Archive for Rational Mechanics and Analysis 219 (3), 1305–1341.

dell'Isola, F., Seppecher, P., Della Corte, A., 2015. The postulations á la D'Alembert and á la Cauchy for higher gradient continuum theories are equivalent: a review of existing results. Proceedings of the Royal Society A 471 (2183), 1–25.

Dunn, J. E., Serrin, J., 1986. On the thermomechanics of interstitial working. In: Dafermos, C. M., Joseph, D. D., Leslie, F. M. (Eds.), The Breadth and Depth of Continuum Mechanics. Springer, Berlin.

Edelen, D. G., Laws, N., 1971. On the thermodynamics of systems with nonlocality. Archive for Rational Mechanics and Analysis 43 (1), 24–35.

Erdle, H., Böhlke, T., 2017. A gradient crystal plasticity theory for large deformations with a discontinuous accumulated plastic slip. Computational Mechanics 60 (6), 923–942.

Ericksen, J. L., 1961. Conservation laws for liquid crystals. Transactions of the Society of Rheology 5 (1), 23–34.

Eringen, A. C., 1964. Simple microfluids. International Journal of Engineering Science 2 (2), 205–217.

Eringen, A. C., 1968. Mechanics of Micromorphic Continua. In: Kröner, E. (Ed.), Mechanics of Generalized Continua. Springer, Berlin.

Eringen, A. C., 1970. Balance laws of micromorphic mechanics. International Journal of Engineering Science 8 (10), 819–828.

Eringen, A. C., 1977a. Edge dislocation in nonlocal elasticity. International Journal of Engineering Science 15 (3), 177–183.

Eringen, A. C., 1977b. Screw dislocation in non-local elasticity. Journal of Physics D: Applied Physics 10 (5), 671–678.

Eringen, A. C., 1992. Balance laws of micromorphic continua revisited. International Journal of Engineering Science 30 (6), 805–810.

Eringen, A. C., 1999. Microcontinuum Field Theories: I. Foundations and Solids. Springer Science & Business Media, New York.

Eringen, A. C., 2002. Nonlocal Continuum Field Theories. Springer, New York.

Eringen, A. C., Suhubi, E. S., 1964. Nonlinear theory of simple micro-elastic solids–I. International Journal of Engineering Science 2 (2), 189–203.

Eugster, S. R., dell'Isola, F., 2017. Exegesis of the Introduction and Sect. I from "Fundamentals of the Mechanics of Continua" by E. Hellinger. Zeitschrift für Angewandte Mathematik und Mechanik 97 (4), 477–506.

Eugster, S. R., dell'Isola, F., 2018a. Exegesis of Sect. II and III.A from "Fundamentals of the Mechanics of Continua" by E. Hellinger. Zeitschrift für Angewandte Mathematik und Mechanik 98 (1), 31–68.

Eugster, S. R., dell'Isola, F., 2018b. Exegesis of Sect. III.B from "Fundamentals of the Mechanics of Continua" by E. Hellinger. Zeitschrift für Angewandte Mathematik und Mechanik 98 (1), 69–105.

Fleck, N., Hutchinson, J., 1997. Strain gradient plasticity. Advances in applied mechanics 33, 295–361.

Forest, S., 2005. Mechanics of Cosserat media–an introduction. Ecole des Mines de Paris, Paris , 1–20.

Forest, S., 2009. Micromorphic approach for gradient elasticity, viscoplasticity, and damage. Journal of Engineering Mechanics 135 (3), 117–131.

Forest, S., 2013. Questioning size effects as predicted by strain gradient plasticity. Journal of the Mechanical Behavior of Materials 22 (3-4), 101–110.

Forest, S., 2019. Micromorphic approach to gradient plasticity and damage. In: Voyiadjis, G. Z. (Ed.), Handbook of Nonlocal Continuum Mechanics for Materials and Structures. Springer, Cham.

Forest, S., Ammar, K., Appolaire, B., 2011. Micromorphic vs. phase-field approaches for gradient viscoplasticity and phase transformations. In: Markert, B. (Ed.), Advances in Extended and Multifield Theories for Continua. Springer, Berlin.

Forest, S., Guéninchault, N., 2013. Inspection of free energy functions in gradient crystal plasticity. Acta Mechanica Sinica 29 (6), 763–772.

Fox, N., 1966. A continuum theory of dislocations for polar elastic materials. The Quarterly Journal of Mechanics and Applied Mathematics 19 (3), 343–355.

Fox, N., 1968. On the continuum theories of dislocations and plasticity. The Quarterly Journal of Mechanics and Applied Mathematics 21 (1), 67–75.

Fried, E., 1996. Continua described by a microstructural field. Zeitschrift für Angewandte Mathematik und Physik 47 (1), 168–175.

Ganghoffer, J., 2007. Differential geometry, least action principles and irreversible processes. Rendiconti del Seminario Matematico Universitá e Politecnico di Torino 65 (2), 43–73.

Germain, N., Besson, J., Feyel, F., 2007. Simulation of laminate composites degradation using mesoscopic non-local damage model and non-local layered shell element. Modelling and Simulation in Materials Science and Engineering 15 (4), 425–434.

Germain, P., 1973. The method of virtual power in continuum mechanics. Part 2: Microstructure. SIAM Journal on Applied Mathematics 25 (3), 556–575.

Giorgio, I., 2016. Numerical identification procedure between a micro-Cauchy model and a macro-second gradient model for planar pantographic structures. Zeitschrift für Angewandte Mathematik und Physik 67 (4), 1–17.

Gottschalk, D., McBride, A., Reddy, B., Javili, A., Wriggers, P., Hirschberger, C., 2016. Computational and theoretical aspects of a grain-boundary model that accounts for grain misorientation and grain-boundary orientation. Computation Materials Science 111, 443–459.

Green, A., 1965. Micro-materials and multipolar continuum mechanics. International Journal of Engineering Science 3 (5), 533–537.

Green, A., Naghdi, P., 1995a. A unified procedure for construction of theories of deformable media. I. Classical continuum physics. Proceedings of the Royal Society of London A: Mathematical, Physical and Engineering Sciences 448 (1934), 335–356.

Green, A., Naghdi, P., 1995b. A unified procedure for construction of theories of deformable media. II. Generalized continua. Proceedings of the Royal Society of London A: Mathematical, Physical and Engineering Sciences 448 (1934), 357–377.

Green, A., Naghdi, P., Rivlin, R., 1965. Directors and multipolar displacements in continuum mechanics. International Journal of Engineering Science 2 (6), 611–620.

Green, A. E., Rivlin, R. S., 1964a. Multipolar continuum mechanics. Archive for Rational Mechanics and Analysis 17 (2), 113–147.

Green, A. E., Rivlin, R. S., 1964b. On Cauchy's equations of motion. Zeitschrift für Angewandte Mathematik und Physik 15 (3), 290–292.

Green, A. E., Rivlin, R. S., 1964c. Simple force and stress multipoles. Archive for Rational Mechanics and Analysis 16 (5), 325–353.

Grmela, M., Öttinger, H. C., 1997. Dynamics and thermodynamics of complex fluids. I. Development of a general formalism. Physical Review E 56 (6), 6620–6632.

Gurtin, M. E., 1993. The dynamics of solid-solid phase transitions 1. coherent interfaces. Archive for Rational Mechanics and Analysis 123 (4), 305–335.

Gurtin, M. E., 1994. The characterization of configurational forces. Archive for Rational Mechanics and Analysis 126 (4), 387–394.

Gurtin, M. E., 1995. The nature of configurational forces. Archive for Rational Mechanics and Analysis 131 (1), 67–100.

Gurtin, M. E., 2002. A gradient theory of single-crystal viscoplasticity that accounts for geometrically necessary dislocations. Journal of the Mechanics and Physics of Solids 50 (1), 5 – 32.

Gurtin, M. E., 2008. A theory of grain boundaries that accounts automatically for grain misorientation and grain-boundary orientation. Journal of the Mechanics and Physics of Solids 56 (2), 640 – 662.

Gurtin, M. E., Anand, L., Lele, S. P., 2007. Gradient single-crystal plasticity with free energy dependent on dislocation densities. Journal of the Mechanics and Physics of Solids 55 (9), 1853–1878.

Gurtin, M. E., Fried, E., Anand, L., 2010. The Mechanics and Thermodynamics of Continua. Cambridge University Press, Cambridge.

Gurtin, M. E., Needleman, A., 2005. Boundary conditions in small-deformation, single-crystal plasticity that account for the burgers vector. Journal of the Mechanics and Physics of Solids 53 (1), 1 – 31.

Hall, E. O., 1951. The deformation and ageing of mild steel: III discussion of results. Proceedings of the Physical Society. Section B 64 (9), 747–753.

Hellinger, E., 1913. Die allgemeinen Ansätze der Mechanik der Kontinua. Encyclopädie der Mathematischen Wissenschaften 4 (4), 601–694.

Hill, R., 1968. On constitutive inequalities for simple materials–I. Journal of the Mechanics and Physics of Solids 16 (4), 229–242.

Hochrainer, T., 2016. Thermodynamically consistent continuum dislocation dynamics. Journal of the Mechanics and Physics of Solids 88, 12–22.

Holzapfel, A. G., 2000. Nonlinear Solid Mechanics II. John Wiley & Sons, New York.

Hütter, G., 2016. An extended Coleman–Noll procedure for generalized continuum theories. Continuum Mechanics and Thermodynamics 28 (6), 1935–1941.

Hütter, M., Svendsen, B., 2011. On the formulation of continuum thermodynamic models for solids as general equations for non-equilibrium reversible-irreversible coupling. Journal of Elasticity 104 (1), 357–368.

Hütter, M., Svendsen, B., 2013. Quasi-linear versus potential-based formulations of force–flux relations and the generic for irreversible processes: comparisons and examples. Continuum Mechanics and Thermodynamics 25 (6), 803–816.

Itskov, M., 2015. Tensor Algebra and Tensor Analysis for Engineers: With Applications to Continuum Mechanics. Springer, Cham.

Kafadar, C., Eringen, A., 1971. Micropolar media–I the classical theory. International Journal of Engineering Science 9 (3), 271 – 305.

Kanso, E., Arroyo, M., Tong, Y., Yavari, A., Marsden, J. G., Desbrun, M., 2007. On the geometric character of stress in continuum mechanics. Zeitschrift für Angewandte Mathematik und Physik 58 (5), 843–856.

Kocks, U., 1976. Laws for work-hardening and low-temperature creep. Journal of Engineering Materials and Technology 98 (1), 76–85.

Kondo, K., 1955. Geometry of elastic deformation and incompatibility. Memoirs of the unifying study of the basic problems in engineering science by means of geometry 1, 5–17.

Krawietz, A., 1986. Materialtheorie. Springer, Berlin.

Krawietz, A., 2015. Classical Mechanics recast with Mach's Principle. Technische Mechanik 35 (1), 49–59.

Lanczos, C., 1949. The Variational Principles of Mechanics. University of Toronto Press, Toronto.

Landau, L., Lifshitz, E., 1969. Mechanics. Pergamon Press, Oxford.

Leslie, F. M., Jan 1968. Some constitutive equations for liquid crystals. Archive for Rational Mechanics and Analysis 28 (4), 265–283.

Liu, I.-S., 1972. Method of Lagrange multipliers for exploitation of the entropy principle. Archive for Rational Mechanics and Analysis 46 (2), 131–148.

Mariano, P. M., 2016. Trends and challenges in the mechanics of complex materials: a view. Philosophical Transactions of the Royal Society A 374 (2066), 1–31.

Marsden, J. E., Hughes, T. J. R., 1994. Mathematical Foundations of Elasticity. Dover, New York.

Maugin, G. A., 1980. The method of virtual power in continuum mechanics: Application to coupled fields. Acta Mechanica 35 (1), 1–70.

Maugin, G. A., 1992. The Thermomechanics of Plasticity and Fracture. Cambridge University Press, Cambride.

Maugin, G. A., 2015. The saga of internal variables of state in continuum thermo-mechanics (1893–2013). Mechanics Research Communications 69, 79–86.

Maugin, G. A., 2017. Non-Classical Continuum Mechanics: A Dictionary. Springer, Singapore.

Mindlin, R., Tiersten, H., 1962. Effects of couple-stresses in linear elasticity. Archive for Rational Mechanics and Analysis 11 (1), 415–448.

Mindlin, R. D., Jan 1964. Micro-structure in linear elasticity. Archive for Rational Mechanics and Analysis 16 (1), 51–78.

Mises, R. v., 1928. Mechanik der plastischen Formänderung von Kristallen. ZAMM-Journal of Applied Mathematics and Mechanics 8 (3), 161–185.

Misra, A., Placidi, L., Scerrato, D., 2017. A review of presentations and discussions of the workshop "Computational Mechanics of Generalized Continua and Applications to Materials with Microstructure" that was held in Catania 29–31 October 2015. Mathematics and Mechanics of Solids 22 (9), 1891–1904.

Moeckel, G. P., 1975. Thermodynamics of an interface. Archive for Rational Mechanics and Analysis 57 (3), 255–280.

Müller, I., 1985. Thermodynamics. Pitman, Boston.

Muschik, W., E. H., 1996. An amendment to the second law. Journal of Non-Equilibrium Thermodynamics 21 (2), 175–192.

Naghdi, P., Srinivasa, A., 1993. A dynamical theory of structured solids. I Basic developments. Philosophical Transactions of the Royal Society A 345 (1677), 425–458.

Neff, P., Ghiba, I.-D., Madeo, A., Placidi, L., Rosi, G., 2014. A unifying perspective: the relaxed linear micromorphic continuum. Continuum Mechanics and Thermodynamics 26 (5), 639–681.

Noether, E., 1971. Invariant variation problems. Transport Theory and Statistical Physics 1 (3), 186–207.

Noll, W., 1958. A mathematical theory of the mechanical behavior of continuous media. Archive for Rational Mechanics and Analysis 2 (1), 197–226.

Orowan, E., 1934a. Zur Kristallplastizität. I. Zeitschrift für Physik 89 (9), 605–613.

Orowan, E., 1934b. Zur Kristallplastizität. II. Zeitschrift für Physik 89 (9-10), 614–633.

Orowan, E., 1934c. Zur Kristallplastizität. III. Zeitschrift für Physik 89 (9), 634–659.

Orowan, E., 1935a. Zur Kristallplastizität. IV. Zeitschrift für Physik 97 (9), 573–595.

Orowan, E., 1935b. Zur Kristallplastizität. V. Zeitschrift für Physik 98 (5), 382–387.

Ortiz, M., Repetto, E., 1999. Nonconvex energy minimization and dislocation structures in ductile single crystals. Journal of the Mechanics and Physics of Solids 47 (2), 397–462.

Öttinger, H. C., Grmela, M., 1997. Dynamics and thermodynamics of complex fluids. II. Illustrations of a general formalism. Physical Review E 56 (6), 6633–6655.

Özdemir, I., Yalcinkaya, T., 2014. Modeling of dislocation–grain boundary interactions in a strain gradient crystal plasticity framework. Computational Mechanics 54 (2), 255–268.

Panoskaltsis, V., Polymenakos, L., Soldatos, D., 2013. A finite strain model of combined viscoplasticity and rate-independent plasticity without a yield surface. Acta Mechanica 224 (9), 2107–2125.

Panoskaltsis, V. P., Soldatos, D., 2014. On spatial covariance, second law of thermodynamics and configurational forces in continua. Entropy 16 (6), 3234–3256.

Peerlings, R., Massart, T., Geers, M., 2004. A thermodynamically motivated implicit gradient damage framework and its application to brick masonry cracking. Computer methods in applied mechanics and engineering 193 (30), 3403–3417.

Petch, N. J., 1953. The cleavage strength of polycrystals. Journal of the Iron and Steel Institute 174, 25–28.

Placidi, L., 2016. A variational approach for a nonlinear one-dimensional damage-elasto-plastic second-gradient continuum model. Continuum Mechanics and Thermodynamics 28 (1), 119–137.

Placidi, L., Barchiesi, E., 2018. Energy approach to brittle fracture in strain-gradient modelling. Proceedings of the Royal Society A 474 (2210), 1–19.

Placidi, L., Barchiesi, E., Misra, A., 2018a. A strain gradient variational approach to damage: a comparison with damage gradient models and numerical results. Mathematics and Mechanics of Complex Systems 6 (2), 77–100.

Placidi, L., Giorgio, I., Della Corte, A., Scerrato, D., 2017. Euromech 563 Cisterna di Latina 17–21 March 2014 Generalized continua and their applications to the design of composites and metamaterials: A review of presentations and discussions. Mathematics and Mechanics of Solids 22 (2), 144–157.

Placidi, L., Misra, A., Barchiesi, E., 2018b. Two-dimensional strain gradient damage modeling: a variational approach. Zeitschrift für Angewandte Mathematik und Physik 69 (3), 1–19.

Planck, M., 1960. A Survey of Physical Theory. Dover, New York.

Polanyi, M., 1934. Über eine Art Gitterstörung, die einen Kristall plastisch machen könnte. Zeitschrift für Physik 89 (9), 660–664.

Prahs, A., Böhlke, T., 2019a. On interface conditions on a material singular surface. Continuum Mechanics and Thermodynamics 32, 1417–1434.

Prahs, A., Böhlke, T., 2019b. On invariance properties of an extended energy balance. Continuum Mechanics and Thermodynamics 32, 843–859.

Rahali, Y., Giorgio, I., Ganghoffer, J., dell'Isola, F., 2015. Homogenization à la Piola produces second gradient continuum models for linear pantographic lattices. International Journal of Engineering Science 97, 148–172.

Rahaman, M. M., Pathak, A., Roy, D., Reddy, J., 2016. Thermo-visco-plasticity under high strain rates: a micro-inertia driven dynamic flow rule. arXiv preprint arXiv:1601.07306 .

Rahaman, M. M., Roy, P., Roy, D., Reddy, J., 2017. A peridynamic model for plasticity: Micro-inertia based flow rule, entropy equivalence and localization residuals. Computer methods in applied mechanics and engineering 327, 369–391.

Rahouadj, R., Ganghoffer, J.-F., Cunat, C., 2003. A thermodynamic approach with internal variables using Lagrange formalism. Part I: General framework. Mechanics Research Communications 30 (2), 109–117.

Sedov, L., I., 1968a. Models of continuous media with internal degrees of freedom. Journal of Applied Mathematics and Mechanics 32 (5), 803–819.

Sedov, L. I., 1968b. Variational methods of constructing models of continuous media. In: Parkus, H., Sedov, L. I. (Eds.), Irreversible Aspects of Continuum Mechanics and Transfer of Physical Characteristics in Moving Fluids. Springer, Vienna.

Seppecher, P., Alibert, J.-J., dell'Isola, F., 2011. Linear elastic trusses leading to continua with exotic mechanical interactions. Journal of Physics: Conference Series 319, 1–13.

Šilhavý, M., 1997. The Mechanics and Thermodynamics of Continuous Media. Springer, Berlin.

Slattery, J. C., Sagis, L., Oh, E.-S., 2007. Interfacial Transport Phenomena. Springer Science & Business Media, New York.

Spring, K. W., 1986. Euler parameters and the use of quaternion algebra in the manipulation of finite rotations: A review. Mechanism and Machine Theory 21 (5), 365–373.

Struchtrup, H., 2008. What does an ideal wall look like? Continuum Mechanics and Thermodynamics 19 (8), 493–498.

Stumpf, H., Hoppe, U., 1997. The application of tensor algebra on manifolds to nonlinear continuum mechanics–invited survey article. ZAMM-Journal of Applied Mathematics and Mechanics 77 (5), 327–339.

Suhubi, E., Eringen, A. C., 1964. Nonlinear theory of micro-elastic solids–II. International Journal of Engineering Science 2 (4), 389–404.

Svendsen, B., 1999. On the thermodynamics of thermoelastic materials with additional scalar degrees of freedom. Continuum Mechanics and Thermodynamics 11 (4), 247–262.

Svendsen, B., 2001a. Formulation of balance relations and configurational fields for continua with microstructure and moving point defects via invariance. International Journal of Solids and Structures 38 (6), 1183–1200.

Svendsen, B., 2001b. On the continuum modeling of materials with kinematic structure. Acta Mechanica 152 (1), 49–79.

Svendsen, B., 2002. Continuum thermodynamic models for crystal plasticity including the effects of geometrically-necessary dislocations. Journal of the Mechanics and Physics of Solids 50 (6), 1297–1329.

Svendsen, B., 2004. On thermodynamic-and variational-based formulations of models for inelastic continua with internal lengthscales. Computer Methods in Applied Mechanics and Engineering 193 (48), 5429–5452.

Svendsen, B., 2011. Continuum thermodynamic and rate variational formulation of models for extended continua. In: Markert, B. (Ed.), Advances in Extended and Multifield Theories for Continua. Springer, Berlin.

Svendsen, B., Bertram, A., 1999. On frame-indifference and form-invariance in constitutive theory. Acta Mechanica 132 (1), 195–207.

Taylor, G. I., 1934. The mechanism of plastic deformation of crystals. Part I.–Theoretical. Proceedings of the Royal Society of London. Series A 145 (855), 362–387.

Toupin, R. A., 1962. Elastic materials with couple-stresses. Archive for Rational Mechanics and Analysis 11 (1), 385–414.

Toupin, R. A., 1964. Theories of elasticity with couple-stress. Archive for Rational Mechanics and Analysis 17 (2), 85–112.

Triani, V., Cimmelli, V. A., 2012. Interpretation of second law of thermodynamics in the presence of interfaces. Continuum Mechanics and Thermodynamics 24 (2), 165–174.

Truesdell, C., Toupin, R., 1960. The classical field theories. In: Flügge, S. (Ed.), Encyclopedia of Physics. Springer, Berlin.

Ubachs, R., Schreurs, P., Geers, M., 2004. A nonlocal diffuse interface model for microstructure evolution of tin–lead solder. Journal of the Mechanics and Physics of Solids 52 (8), 1763–1792.

Vardoulakis, I., 2019. Cosserat Continuum Mechanics: With Applications to Granular Media. Springer, Cham.

Willner, K., 2003. Kontinuums-und Kontaktmechanik: Synthetische und Analytische Darstellung. Springer, New York.

Wulfinghoff, S., 2014. Numerically efficient gradient crystal plasticity with a grain boundary yield criterion and dislocation-based work-hardening. Doctoral Dissertation, KIT Scientific Publishing, Schriftenreihe Kontinuumsmechanik im Maschinenbau Nr. 5, Karlsruhe.

Wulfinghoff, S., Bayerschen, E., Böhlke, T., 2013. A gradient plasticity grain boundary yield theory. International Journal of Plasticity 51, 33–46.

Wulfinghoff, S., Böhlke, T., 2012. Equivalent plastic strain gradient enhancement of single crystal plasticity: theory and numerics. Proceedings of the Royal Society A 468 (2145), 2682–2703.

Wulfinghoff, S., Forest, S., Böhlke, T., 2015. Strain gradient plasticity modeling of the cyclic behavior of laminate microstructures. Journal of the Mechanics and Physics of Solids 79, 1–20.

Yavari, A., Goriely, A., 2012. Riemann–Cartan geometry of nonlinear dislocation mechanics. Archive for Rational Mechanics and Analysis 205 (1), 59–118.

Yavari, A., Marsden, J. E., 2009. Covariant balance laws in continua with microstructure. Reports on Mathematical Physics 63 (1), 1–42.

Yavari, A., Marsden, J. E., Ortiz, M., 2006. On spatial and material covariant balance laws in elasticity. Journal of Mathematics and Physics 47 (4), 1–53.

Schriftenreihe Kontinuumsmechanik im Maschinenbau Karlsruher Institut für Technologie (KIT) (ISSN 2192-693X)

Herausgeber: Prof. Dr.-Ing. Thomas Böhlke

Band 1 Felix Fritzen
Microstructural modeling and computational homogenization of the physically linear and nonlinear constitutive behavior of micro-heterogeneous materials. 2011
ISBN 978-3-86644-699-1

Band 2 Rumena Tsotsova
Texturbasierte Modellierung anisotroper Fließpotentiale. 2012
ISBN 978-3-86644-764-6

Band 3 Johannes Wippler
Micromechanical finite element simulations of crack propagation in silicon nitride. 2012
ISBN 978-3-86644-818-6

Band 4 Katja Jöchen
Homogenization of the linear and non-linear mechanical behavior of polycrystals. 2013
ISBN 978-3-86644-971-8

Band 5 Stephan Wulfinghoff
Numerically Efficient Gradient Crystal Plasticity with a Grain Boundary Yield Criterion and Dislocation-based Work-Hardening. 2014
ISBN 978-3-7315-0245-6

Band 6 Viktor Müller
Micromechanical modeling of short-fiber reinforced composites. 2016
ISBN 978-3-7315-0454-2

Die Bände sind unter www.ksp.kit.edu als PDF frei verfügbar oder als Druckausgabe bestellbar.

Band 7 Florian Rieger
Work-hardening of dual-phase steel. 2016
ISBN 978-3-7315-0513-6

Band 8 Vedran Glavas
Micromechanical Modeling and Simulation of Forming Processes. 2017
ISBN 978-3-7315-0602-7

Band 9 Eric Bayerschen
Single-crystal gradient plasticity with an accumulated plastic slip: Theory and applications. 2017
ISBN 978-3-7315-0606-5

Band 10 Bartholomäus Brylka
Charakterisierung und Modellierung der Steifigkeit von langfaserverstärktem Polypropylen. 2017
ISBN 978-3-7315-0680-5

Band 11 Rudolf Neumann
Two-Scale Thermomechanical Simulation of Hot Stamping. 2017
ISBN 978-3-7315-0714-7

Band 12 Mauricio Lobos Fernández
Homogenization and materials design of mechanical properties of textured materials based on zeroth-, first- and second-order bounds of linear behavior. 2018
ISBN 978-3-7315-0770-3

Band 13 Malte Schemmann
Biaxial Characterization and Mean-field Based Damage Modeling of Sheet Molding Compound Composites. 2018
ISBN 978-3-7315-0818-2

Band 14 Jürgen Albiez
Finite element simulation of dislocation based plasticity and diffusion in multiphase materials at high temperature. 2019
ISBN 978-3-7315-0918-9

Band 15 Maria Loredana Kehrer
Thermomechanical Mean-Field Modeling and Experimental Characterization of Long Fiber-Reinforced Sheet Molding Compound Composites. 2019
ISBN 978-3-7315-0924-0

Band 16 Peter Hölz
A dynamic and statistical analysis of the temperature- and fatigue behavior of a race power unit – The effect of different thermodynamic states. 2020
ISBN 978-3-7315-0988-2

Band 17 Andreas Prahs
A Gradient Crystal Plasticity Theory Based on an Extended Energy Balance. 2020
ISBN 978-3-7315-1025-3